Human Toxicology
of Pesticides

Authors

Fina P. Kaloyanova

Professor
Head, Department of Toxicology
Research Institute of Hygiene and Occupational Health Medical Academy
Sofia, Bulgaria

and

Mostafa A. El Batawi

Professor
Adjunct, University of Pittsburgh
Graduate School of Public Health;
World Health Organization Consultant
Former Director of Occupational Health
WHO, Geneva, Switzerland

CRC Press
Boca Raton Ann Arbor Boston London

Library of Congress Cataloging-in-Publication Data

Kaloyanova, Fina, P.
    Human toxicology of pesticides / authors, Fina P. Kaloyanova, Mostafa A. El Batawi.
        p. cm.
    Includes bibliographical references and index.
    ISBN 0-8493-5192-8
    1. Pesticides — Toxicology. I. El-Batawi, Mostafa A. II. Title.
RA1270.P4K278 1991
615.9'51—dc20   91-8252
                                                                                CIP

This book represents information obtained from authentic and highly regarded sources. Reprinted material is quoted with permission, and sources are indicated. A wide variety of references are listed. Every reasonable effort has been made to give reliable data and information, but the author and the publisher cannot assume responsibility for the validity of all materials or for the consequences of their use.

Direct all inquiries to CRC Press, Inc., 2000 Corporate Blvd., N.W., Boca Raton, Florida, 33431.

International Standard Book Number 0-8493-5192-8

Library of Congress Card Number 91-8252
Printed in the United States

0 1 2 3 4 5 6 7 8 9

PREFACE

The toxicity of pesticides has generated an extensive amount of literature, because of their domestic and public health use, and particular because of their wide use in agriculture, which employs the largest workforce in the world. This book adds new elements and provides for wider needs in the health profession. It has a thorough review of literature on the subject and results of research held in both Eastern and Western countries, which is rare to find in one volume. The literature cited is one of the largest.

For each group of chemical compounds used as pesticides, the authors desribe all the known diagnoises and therapies; therefore, this book will be used not only by public health personnel and occupational physicians working in the primary health care systems in rural areas, but also by medical care professionals and research institutions. In several parts of the book the authors refer to the gaps in knowledge that require further research.

The chapter on combined effects is expectedly rather short in view of the limited information available on the subject. For more information about combined exposures and effects, the reader may refer to a WHO publication in the Technical Report Series No. 662, in which the mechanisms of response to combined exposure and the variety of effects, e.g., synergism, addition, or independent effects, that may result.

The term "TLV" has been widely used throughout the text, because of its familiarity to most of the readers. It denotes a quantitative hygienic standard for a considered safe level expressed as a concentration with a defined average time. It can also be taken to mean maximum allowable concentration, MAC.

The authors are fully aware of the WHO program on exposure limits, ELs, which should globally replace TLVs. ELs are internationally recommended and they are based on scientific and epidemiological criteria that are available in the world; they are health based. An occupational exposure limit, OEL, signifies that level of exposure, which in the case of a daily work exposure of 8 hours, will not cause in the individual any disease or disorder from a normal state of health during his working or entire life time or in any of his dependents. This is also fully supported by the ILO. The word "threshold" in TLV is taken in Eastern European countries as the beginning of adverse health effects.

Mention of TLV in place of exposure limits was only made for convenience in view of the fact that not many readers are familiar with ELs; however, they generally mean the same thing in practice. It has been internationally agreed that exposure limit is the more accurate term in scientific and health meaning.

F. K. and
M. ELB.

THE AUTHORS

Dr. Fina P. Kaloyanova, M.D., Ph.D., D.Sc., is professor and head of the Department of Toxicology in the Research Institute of Hygiene and Occupational Health, Medical Academy, Sofia, Bulgaria.

Dr. Kaloyanova graduated in 1951 to the Medical Faculty in Sofia. In 1961 she obtained her Ph.D. and in 1974 her D.Sc. in the field of pesticide toxicology. She began working as a researcher in 1951, a senior researcher in 1961, and a professor in 1969. From 1972 to 1984 she was the director of the Research Institute of Hygiene and Occupational Health.

Dr. Kaloyanova was a president of the Bulgarian Society of Preventive Medicine and now is a member of the board of the National Association of Medical Research Societies. She was vice president of the International Association of Rural Medicine from 1972 to 1975, and a board member of the Permanent Commission and Internatinal Association for Occupational Health from 1972 to 1978. She was elected as a member of the International Academy of Environmental Safety.

She has received several Distinguished Service Awards from the Medical Academy, Ministry of Public Health, Bulgarian Government.

Dr. Kaloyanova has been a member of the expert panel on occupational health of the World Health Organization since 1972. She participated as a temporary advisor and consultant of more than 50 WHO meetings. She has presented many guest lectures at international and national courses in Egypt, Cuba, China, U.S.S.R, Algeria, Poland, Yugoslavia, and elsewhere. Dr. Kaloyanova has published over 250 research papers and books. Her current major research interests relate to pathogenesis of the pesticide intoxication and determination of the threshold level of the toxic action.

Dr. Mostafa A. El Batawi, M.D., D.Sc., is an adjunct professor at the University of Pittsburgh Graduate School of Public Health and a consultant to the World Health Organization, Geneva, Switzerland.

Dr. El Batawi received his M.D. in 1951 from Cairo (Egypt) University Faculty of Medicine. In 1957, he earned his Masters of Public Health from the University of Pittsburgh and received his D.Sc. in Occupational Health two years later. From 1957 to 1960 he was a resident in internal medicine, neuropsychiatry, and psychiatry at St. Francis General Hospital, Montifore Hospital, and Western's Psychiatric Institute and Clinic, University of Pittsburgh, respectively.

Dr. El Batawi served as the Chief Medical Officer for the Office of Occupational Health, WHO Headquarters, Geneva, from 1970 to 1988. He has been chairman and professor of the Department of Occupational and Environmental Health Sciences, College of Medicine, New York Institute of Technology. He served as Regional Advisor in Occupational Health for Asia and Western Pacific with the International Labour Organization and the United Nations' Development Programme and was associate professor and head of the Department of Occupational Health, University of Alexandria, Egypt.

Dr. El Batawi has been a member of the International Commission in Occupational Health since 1965 and was a board member from 1981 to 1988. He is a member of the International Epidemiological Association, the American Occupational Medical Association, the American Industrial Hygiene Association, and the Society for Advanced Medical Systems. He is an honorary member of the German Association of Industrial Medicine, the Polish Association for Occupational Medicine, and the International Association for Agricultural Medicine and Rural Health. Dr. El Batawi is also a member of the American College of Preventive Medicine, the Scientific Committee on Occupational Epidemiology of the International Commission on Occupational Health, the New York Academy of Sciences, and the American Human Factors Society.

Dr. El Batawi received the third Theodore F. Hatch Lecture and award from the Graduate School of Public Health, Pittsburgh, and the American Conference of Governmental Industrial Hygienists, as well as the Distinguished Graduate Medallion from the Graduate School of Public Health's Alumni Association, University of Pittsburgh. He was awarded the Shield of Honor for Serving Humanity in the Fields of Medicine and Health from the Union of Medical Professions, Cairo, Egypt. He has also received the Award of the National Institute for Occupational Safety and Health of the United States for Services to the Health of Working People in the World and the Commander's Cross for "the best contribution to Occupational Health" from the President of the Republic of Poland.

CONTENTS

Chapter 1

INTRODUCTION

Many years before our era, different natural substances were used as pesticides. Later, salts of metals, sulfur, natural oils, and tobacco products were applied. During the last 50 years chemical synthesis of pesticides has increased considerably, and nowadays there are more than 55 classes and 1500 individual substances produced in more than 100,000 formulations.

According to the definition of the FAO International Code of Conduct on the Distribution and Use of Pesticides, a "pesticide is any substance or mixture of substances intended for preventing, destroying or controlling any pest, including vectors of human or animal disease, unwanted species of plants or animals causing harm during or otherwise interfering with production, processing, storage, transport or marketing of food, agricultural commodities, wood and wooden products or animal feedstuffs, or which may be administered to animals for the control of insects, arachnids or other pests in or on their bodies. The term includes substances intended for use as a plant-growth regulator, defoliant, dessicant, or fruit tinning agent for preventing of premature fall of fruit and substances applied to crops either before or after harvest to protect the commodity from deterioration during storage and transport."

According to their target, pesticides are divided into several main groups:

1. Insecticides
2. Other insect-control agents
        Chemosterilants
        Pheromons (sex attractants and synthetic lures)
        Repellents
        Insect hormones and hormone mimics (insect growth regulators)
3. Specific acaricides
4. Protectant fungicides
5. Eradicant fungicides (chemotherapeutants)
6. Soil fumigants and nematocides
7. Herbicides
8. Dessicants, defoliants, and haulm killers
9. Rodenticides
10. Plant growth regulators
11. Molluscicides

The specific conditions of pesticide application in agriculture, forestry, industry, public health, and households make them one of the most common type of chemicals coming into contact with all groups of a population. They have gained a widespread use in all countries due to their proven effect in vector control and their high effectiveness in agriculture.

However, pesticides represent a very serious health and environmental problem; preventing their eventual adverse effects is much more difficult than is the case with other substances used in industry.

Pesticide application mainly requires scattering onto large areas (millions of hectares of soil) and using concentrations capable of killing selected plant and animal species. It is evident that the protective measures for both workers and useful animal species are very limited and specific. Their circulation in the environment cannot be brought under control. Therefore, public and occupational circles are greatly concerned with coping with the present and eventual unknown future and the delayed adverse effects of pesticides that represent a hazard for human progeny and the environment. The hopes are set on integrated pest management.

The present book only concerns itself with the adverse effects pesticides have on human health. The book attempts to evaluate the available literature data, as well as the authors' own investigations on the human toxicology of pesticides. The data on some groups of compounds are very poor, on others nonexistent.

We hope the book will be useful, not only to physicians, but to specialists in different fields who are interested in the problems related to pesticides.

Chapter 2

ORGANOPHOSPHOROUS COMPOUNDS

I. INTRODUCTION

The largest group of pesticides used nowadays is the group of organophosphorous compounds (OP). More than 100 individual compounds of this group are well known and largely used in many countries.

OP degradation in the environment, compared with organochlorine pesticides (OCP), is very fast. Their stability in the environmental media is calculated in months; OCPs persist for years in the environment. In many cases, OP pesticides were a good alternative to previous application of OCPs, especially DDT and hexachloran, as well as, to some extent, some pesticides from the cyclodien group. A disadvantage of OP pesticides is their high acute toxicity and the resulting large number of fatal acute intoxications.

Synthesis of OP compounds is still in progress, and substances with more specific or selective effects have been synthesized. At the same time, they are less toxic for higher organisms, including man.

II. PROPERTIES

A. CHEMICAL STRUCTURE
Table 1 presents the general chemical structures of the principal OP groups together with the common or other pesticide names from each group. An extensive list of OPs with trade names is given in *WHO EHC No 63,*[1] where molecular weight, CAS registry number and CAS chemical name are given as well. Molecular weight of OPs varies from 183 to 466.

They are formulated as water soluble powder, liquid concentrate, or granules. All are rapidly hydrolyzed and oxydized, in the environment and in alkali media, to mono- or disubstituted phosphoric or phosphonic acid or their thioanalogs. Some isomerization reactions occur while storing OPs at high temperatures, and more toxic derivatives are formed. Humidity and sunlight also play a role in the transformation of OPs under natural conditions. Degradation in the environment involves both hydrolysis and oxidation to water soluble products. Microbial degradation also contributes to the fast disappearance of OPs from the treated fields.

III. USES

OPs are the most used group of pesticides worldwide in view of their efficiency and rapid degradation in the environment and living organisms. Most are insecticides, but some are used as fungicides, herbicides, and raticides. They are used in agriculture and forestry and for public health purposes as a 0.02 to 0.08% water solution for spraying, 2.5 to 8% as aerosols and 4% as dust.[2]

IV. METABOLISM

OPs enter the body through the nose, skin, or mouth. Uptake through the skin may be very extensive for OPs, since more of them are lipophillic. Percutaneous penetration of man by parathion and malathion labeled with radioactive carbon was studied by Maibach and Feldmann.[3] Pesticides were applied on different anatomic regions of volunteers. As a criterion for absorption, urinary recovery of ^{14}C was studied. The scrotum allowed almost total absorption of

3

TABLE 1
General Chemical Structure and Common or Other Names of Ogranophosphorous Insecticudes

Type of phosphorus group	Outline structure	Common or other names
Phosphate	O ‖ (R—O)$_2$—P—O—X	Chlorfenvinphos, crotoxyphos, dichlorphos, dicrotophos, heptenphos, mevinphos, monocrotophos, naled, phosphamidon, TEPP, tetrachlorvinphos, triazophos
O-alkyl phosphorothioate	O ‖ (R—O)$_2$—P—S—X	Amiton, demeton-S-methyl, omethoate, oxydemetopmethyl, phoxim, vamidothion
	S ‖ (R—O)$_2$—P—O—X	Axothoate, bromophos, bromophos-ethyl, chlorpyriphos, chlorpyriphos-methyl, coumaphos, diazinon, dichlofenthion, fenchlorphos, fenitrothion, fenthion, fensulphothion, iodofenphos, parathion, parathion-methyl, phoxim, pyrasophos, pirimiphos-methyl, sulfotep, temephos, thionazin
Phosphorodithioate	S ‖ (R—O)$_2$—P—S—X	Amidithion, azinophos-ethyl, azinophos-methyl, dimetoate, dioxathion, disulfoton, ethion, formothion, malathion, mecarbam, menazone, methidathion, morphothion, phentoate, phorate, phosalon, phosmet, prothoate, thimeton

S-alkyl phosphorothioate	R
	\S O
	  \ ‖
	   P—O—X
	  /
	O
	/
	R
		Profenofos, trifenofos

S-alkyl phosphorodithioate	S R—S ‖ \ P—O—X R—O /	Prothiofos, sulprofos
Phosphoramidate	O ‖ (R—O)$_2$—P—NR$_2$	Cruformate, fenamifos, fosthietan

Phosphorotriamidate	O
	‖
	R$_2$N—P—N
	      |
	      NR$_2$
		Triamifos

Phosphorothioamidate	O
	‖
	R—O—P—NR$_2$
	     |
	     S—alkyl
		Methamidofos (tamaron)

	S ‖ (R—O)$_2$—P—NR$_2$	Isofenfos

TABLE 1 (continued)
General Chemical Structure and Common or Other Names of Ogranophosphorous Insecticudes

Type of phosphorus group	Outline structure	Common or other names

Phosphonate	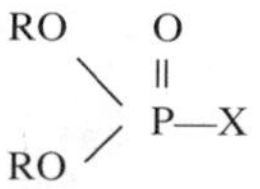 RO O
	  \\ ‖
	    />P—X
	RO
		Butonate, trichlorfon (tribyton)

Phosphonothioate	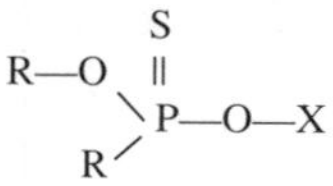 S R—O ‖ \\>P—O—X R	EPN, Trichlormat, leptophos, cyanofenphos

parathion, axilla — 63%, ear canal — 46%, forehand — 36%, scalp — 32%, foot — 14%, palm — 11%, forearm — 8.6%. Malathion penetration was lower.

A. BIOTRANSFORMATION

The three main reactions for OP biotransformation are biochemical oxidation, hydrolysis, and transferase reaction.

B. BIOCHEMICAL OXIDATION

Mixed function oxidases (MFO) are involved in this process. Many OPs can be oxidized by MFOs in liver endoplasmic reticulum, but also by MFOs found in the intestines, kidney, and lung.

Oxidative desulfuration of P=S groups to P=0 almost always leads to more toxic products. Phosphorothioate activates to phosphate, a direct ChE inhibitor. Oxones are less lipophilic and more rapidly hydrolyzed. This reduces their accumulation in the body.

Other reactions mediated by MFOs are oxidative *N*-dealkylation, oxidative *O*-dealkylation, oxidative dearylation, thioeter oxidation, and side-chain oxidation.

C. HYDROLYSIS

In OP hydrolysis, some hydrolytic enzymes are involved. Commonly known among them are A-esterases or phosphoryl phosphatases. Some authors call them in accordance with the OP compound used as a substrate — "paraoxonase", "malaoxonase", etc. Georgiev has recently given a scheme for the biotransformation of OPs (Figure 1).[5]

V. TOXICITY: MECHANISM OF ACTION

OPs show identical toxic effects, mainly due to the cholinesterase inhibition. Three main biochemical reactions are responsible for the effect of OPs.[6,7]

1. Inhibition of cholinesterase activity
2. Inhibition of neuropathy target esterase (NTE) and development of delayed neuropathy
3. Release of alkyl groups attached to the phosphorous atom and alkylation of macromolecules including RNA and DNA

The principal mechanism of OP action is inhibition of cholinesterase activity(ChEA), the enzyme performing the hydrolysis of acetylcholine to choline and acetic acid. Specific ChE(acetylcholinesterase 3.1.1.7) is located in the nervous ganglionic synapses of neuromus-

Activation
P = S ⇌ P = O

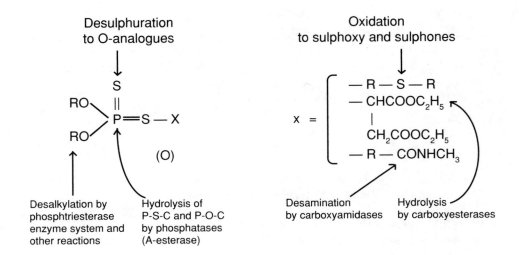

FIGURE 1. The general scheme for the main metabolite process of OPs (adapted by Georgiev[5]).

cular structures and in erythrocytes. Nonspecific ChE(3.1.1.8) is found mainly in the plasma and liver. Organophosphorous insecticides generally inhibit both enzymes.

With the change of the membrane potential, acetylcholine (ACh) acts as a mediator of the nerve impulse. In the cytoplasm of the nerve end, before the synaptic membrane, special vesicles contain acetylcholine, which is synthesized by the enzyme cholinacetylase (ChA) from acetyl CoA and choline.

The nerve impulse produces a discharge of ACh across the synaptic gap. The ACh contacts cholinergic receptor protein molecules of the postsynaptic membrane and changes its configuration, enabling Na and K cations to penetrate. The transfer of the nerve impulse continues. This process is very short, lasting about 1/500 s, and is followed by ChE hydrolysis of ACh. In normal conditions the half-life for the hydrolysis of acetylted ChE is 2.3×10^{-6} min.

The OPs produce phosphorylation of the ChE esteric binding site. The half-life of dimethyl posphorylated ChE hydrolysis is about 50 min, and for diethyl phosphorylated it is about 60 h. Inactivation by ChE phosphorylation stops the hydrolysis of ACh. Excessive quantities of ACh accumulate at peripheral ganglionic and central nerve endings (synapses) in effector organs, elevated concentrations occur in plasma and intestinal fluid. The intoxication effects connected with the excitement of *M*- and *N*-choline receptors (present on nerve terminals of effector organs) are as follows:

- Muscarine effect due to postganglionic cholinergic nerve impulses exciting the *M*-choline receptors of the lungs, gastrointestinal system, heart, kidneys, sweat glands, pupils and muscles
- Nicotinic effect on the receptors of ganglionic synapses and motoric plates, the medular part of glandula subrenalis, and carotic nodules

Central effect of ACh due to nerve cells or ACh accumulation directly impacting the choline receptors, with parallel inhibition of other enzymes by OPs such as lipase, cholesterol esterase, proteinase, monoaminooxidase, and other nonspecific esterases

The duration of the symptoms depends partly on the rate of ChE reactivation. Spontaneous reactivation depends on the chemical structure of the phosphoryl group attached to the enzyme. The reactivation of the inhibited enzyme can be facilitated considerably by special compounds (oxymes). Several of these compounds have become important antidotes in the treatment of poisoning.

The inhibited enzyme may also be transformed into a state where no spontaneous reactivation occurs and where oximes are no longer capable of reactivating it. The phenomenon is called "aging" and is characterized by removal of one of the alkyl groups from the phosphoryl groups attached to the enzymes. The rapidity of inhibited ChE aging depends on the chemical nature of the phosphorylating insecticides.[9,10]

OPs inhibit a class of esterases which use serin in their catalytic centre and are called serin esterases or *B*-esterases.[8] This inhibition is not significant for acute toxicity, but it may have other toxicological significance (potentation, detoxication, etc.).

The inhibition of neuropathy target esterase (NTE) protein, found in the nervous system, is responsible for the development of dalayed neupathy.[11]

Delayed neuropathy can only occur for some phosphates and phosphonates — i.e., one of the groups (R_1 or R_2) is attached to the phosphorus atom by oxygen or nitrogen. Phosphonates, in which both R_1 and R_2 are attached directly to the phosphorous atom by P-C bonds, do not cause delayed neuropathy. Although they may be potent inhibitors of neurotoxic esterase, the inhibited enzyme cannot age, and therefore, does not cause delayed neuropathy.

Lotti and Johnson performed comparative studies on the distribution of NTE and ChE in brain.[12] ChEA was greater in the nucleous caudatus, but the NTE was found in almost the entire cortex area as well as in some parts of the cerebellum and the spinal cord. These authors also detected NTE in the femoral nerve — 50 nM/min/g wet weight. The normal values of NTE in the human cortex are 2,390 + 100 nM/min/g wet weight.

VI. EXPOSURE: DOSE-EFFECT RELATIONSHIP

If good agricultural practice is followed, the exposure of general population by food is negligible. The only possible serious exposure could occur during OP spraying for public health or for individual oral or local application against some parasitic diseases. Oral dose-effect relationship was studied during treatment of some parasitic diseases and in volunteer studies (Table 2).

Dipterex was used to treat schistosomiases in thirty children from 4 to 12 years old, in doses of 5 to 10 mg/kg in relation to age, for 10 d. The inhibition of erythrocyte ChE changed from 19 to 52%, with a mean of 32%. Pseudocholinesterase decreased insignificantly, by maximum 15%. No liver, hematopoetic, or renal impairments were demonstrated.[17]

Similar results are reported by Beneyt et al.[14] They used a dose of 500 mg dipterex to treat adults twice in two consecutive days with no adverse effect. Only ChE was inhibited. Other patients, treated with 1 g daily and 1.3 g daily, suffered to a greater or lesser degree of gastrointestinal pains. The ChE inhibition was considerable, in some cases reaching less than 20% the normal mean level. In this study the inhibition of ChE activity was pronounced.

A. OCCUPATIONAL EXPOSURE

Occupational exposure to OPs has been the subject of many studies concerned with preventing eventual health impairments. Data concerning exposure concentrations and OP effects are shown in Table 3.

TABLE 2
Oral Dose-Effect Relationship In Some OPs[15,16]

Name of Pesticide	Dose in mg/kg with no observed effect	Dose with observed effect in mg/kg/b.w.	
		Dose/term of application	Kind of effect
Azynophosmethyl	0.2/30 d	0.3–0.33 (time undetermined)	ChE inhibition
Abate	256 mg/man/d 5 d 62 mg/kg/d 4 weeks		
Bromophos	0.4/4 weeks		
Dichlorophos	0.033/28 d	1/d 3 weeks	Inhibition of ChE plasma
		8/d 3 weeks	Inhibition of ChE plasma by 70%, gastrointestinal disorders
		16/d 5 d	Inhibition of ChE plasma by 90%, inhibition of erythrocyte ChE
		5/d 10–20 d	Inhibition of ChE plasma by 25%, no inhibition of Er ChEA
Demethon *S*-methyl	0.5/single	0.4/30 d	ChE inhibition by 20%
Diazinon	0.02/34 d	0.25/43 d	Plasma ChE inhibition by 20%
		0.05/5 d, repeated after 23 d	Plasma ChE inhibition by 35%
Dimethoate	0.2/57 d	0.5/single	ChE inhibition
Dioxathin	0.075/28 d		
Disulfotan	0.075/30 d		
Chloropyriphos	0.1/1 month		
Fenchlorophos	10.7/4 d	19.4/7 d	ChE inhibition
Formothion	0.2/59 d	0.5/59 d	ChE inhibition
Malathion	0.2/88 d	0.4/88 d	ChE inhibition
		0.8/2 weeks	ChE inhibition
Methidathion	0.11/6 weeks		
Mevinphos	0.014/30 d	0.025/single	Inhibition of erythrocyte ChE only
Trichlorfon		7.5/one dose at 2 weeks intervals	ChE inhibition nausea
		7/four time at fortnight intervals	Diarhea, abdominal pain, restlessness
		24/single	Tachycardia, abdominal pain, vomiting, tremor, sweating
Parathion	0.05/3 weeks	0.12/6 weeks	Plasma and erythrocyte ChE inhibition by 35%
	0.06/43 d	0.1/28 d	Plasma ChE inhibition by 15%
		0.12/27 d	
Phenitrothion	0.04–0.08 × 24 h intervals	0.3/single dose	ChE inhibition
Phenthion	0.02/single		
Pyrimiphosmethyl	0.25/28 d		

TABLE 3
Occupational Exposure to OP Concentration-Effect Relationship

Pesticides	Concentration (mg/m³)	Symptoms
Dichlorphos[15]	8/21 d	Mild symptoms
	1/21 d	ChE inhibition
	0.033/28 d	No effects
	6.9/30 to 60 min	No symptoms and no ChE inhibition
	0.9 to 3.5/8 h	Slight ChE inhibition
	0.25/11 weeks, 4 d a week	No changes
	0.51/single	Slight reduction of ChE; tolerated dose received by inhalation is about 0.5 mg/man/d, 1 mg/man/d produced slight plasma ChE reduction
Bromophos[15]	During spraying for 14 d	Slight reduction of ChE max by 25% — reactivation in a month
Fenitrothion[15]	One week spraying	Inhibition of blood ChE by 50% in some of the sprayers
Shradan[18]	0.1–3	Blood ChE inhibition, constriction of pupils
	0.03–0.9 (duration of exposure not specified)	Burning eyes, headache, weakness
Malathion[19]	0.1–3 long-term exposure	Inhibition of ChE in 1/2 of the workers, decrease of albumin and beta and gama globulines
Parathion[20]	0.1–0.8/short-term exposure	ChE reduction
	6–13/short-term exposure	Acute intoxication
DDVP[21]	0.8–3/1 d	ChE inhibition by 19–23%, no symptoms
Trichlorphor[20]	0.010–0.034/short-term spray operators	No changes
Phosfolan[23]	0.008–0.3% toxic dose	Inhibition of erythrocyte ChE by 31–44%; partial recovery in 48 h, full recovery in 3–4 weeks
Dipterex[24]	1.54, 0.38, 0.52 long-term exposure in formulation plant	Inhibition of blood ChE, EEG changes

Wolfe et al. studied the magnitude of potential dose to be absorbed during exposure to the air blasted in power drawn tractors.[25] Values for potential dermal and respiratory exposure, and for total exposure in terms of fractions of toxic dose, were determined for different pesticides during orchard spraying. The highest total exposure was calculated to be only 1.12% of the toxic dose/h for workers applying the OP compound carbophenothion, the most toxic compound in the study. In another study, an orchard exposed to parathion was calculated to be 19 mg/h dermal and 0.02 mg/h inhalatory[2]. In practical conditions workers wear the required protective gear, which considerably reduces the chemical absorption. Wolfe et al. calculated potential parathion exposure in a parathion formulating plant based on minimum protective means (without respirator or gloves and with short sleeves).[26] The mean dermal exposure for 11 workers was 67.3 mg/h, and the mean respiratory level was 0.62 mg/h; this represents 5% of the toxic dose per h. The highest values obtained during the test period were 335 mg/h dermal and 3.4 mg/h respiratory exposure, equal to 25% of the toxic dose. Such an exposure represents a big hazard.

Cavanga et al. and Cavanga and Vigliani studied the effects of vapona strips in a hospital. They determined the ChEA in persons exposed to different vapona concentrations in the air for about 11 d. In 66 patients they only found inhibition of plasma ChEA levels ranging from 35 to 72% (average 54%) when exposure concentrations were 0.1 to 0.28 mg/m³. No changes were found at concentrations of 0.02 to 0.1 mg/m³, except for patients with liver disease. The authors

TABLE 4
Potential Route of Intoxication by Demethon and Parathion[29]

Route of absorption	%
Dermal	71
Inhalatory	8
Oral	11
Miscellaneous	10

calculated the intake of vapona, assuming complete absorption and air 10 m³/d inhaled air in adults and children. They found that 1.73 mg vapona daily is the average absorption in adults, which provokes 54% inhibition of ChE, 0.34 mg/daily for liver patients — 44% inhibition, and 0.2 mg/daily for children — 25% inhibition. No symptoms were recorded.

A study by Shih et al. suggested that a percutaneous route of entry was the dominant one among sprayers.[29] They analyzed the reasons for parathion and demethon intoxications in an area of China and found 71% to be related to dermal absorption (no protective gloves, contaminated clothes, unclothed arms and legs, leaky tools, and spraying against the wind). The distribution of the intoxications by route of absorption is shown in Table 4.

Direct evidence for their conclusion are the low concentrations measured in the breathing zone (0.02 mg/m³ in only one of the samples) and the high amount of parathion contamination on the skin, e.g., arm (30 cm²) — 5.7 mg, glove or right hand — 1.271 mg, and leg (30cm²) — 30.8 mg.

Dermal absorption has been proved to be the primary route of malathion intake during spraying and mixing operations.[30] The average daily dermal exposure has been 330 mg and the absorbed dose was calculated to have been 26 mg daily. Respiratory exposure to airborn malathion was extremely low (mean 0.43 mg/m³ and peak 1.54 mg/m³).

Pesticide spraying has been shown to deposit on exposed surface 20 to 1700 times more than the amount reaching the respiratory tract .[31]

Davies et al. undertook a study to determine the potential dermal and respiratory exposures experienced apple thinners received while working in orchards 24 and 48 h after the first seasonal cover spray of phosalone.[32] The total foliar residues found in the study were 2.1 μg/cm² on both days. The potential dermal and respiratory exposure were reported to be, respectively, 9.0 and 0.13 mg/h at 24 h and 7.1 and 0.081 mg/h at 48 h. The thinners were assumed to have received 0.034% of a lethal dose by the dermal route and 0.0015% of the lethal dose/h by the respiratory route. While working their normal 7.5 h a day, they would receive a total of 0.27% of the lethal dose.

Farmers planting corn and applying granulated terbufos as a soil insecticide and nematocide were estimated to have dermal exposure ranging from 5 to 156 μg/h, with an average of 72.4 μg/h, and respiratory exposure ranging from 2.8 to 27.4 μg/h, with an average of 11 μg/h. The estimated percent of toxic doses calculated ranged from 0.01 to 0.20 with an average of 0.111. No detectable absorption of terbufos was found in the farmers, as indicated by the results of the urinary alkyl phosphate analyses and cholinesterase measurements. The conception, based on these results, is that the granular form of terbufos does not present a significant hazard. It should be mentioned that the acute LD_{50} of the active ingredient is 1.1 mg/kg by dermal application.[33]

VII. EFFECTS ON HUMANS

A. ACUTE OP INTOXICATION

The number of accidental and occupational poisonings from OP compounds seems to correlate well with their absolute toxicity. According to Namba, during a seven year period in Japan there were 63 cases of malathion intoxication, including 10 deaths, and 3311 cases of parathion intoxication, including 188 deaths.[32] The relatively low potential hazard of malathion

TABLE 5
OP Acute Intoxication Symptoms And Signs (Adapted)[9]

Site of action	Signs and symptoms
Eyes	Increased lacrimation, slight myosis (occasionally unequal, later marked), blurred vision, eye pain when focusing, frontal headache, conjunctive hyperemia
Respiratory system	Rhinorrhea, hyperemia (local exposure), tightness in chest, prolonged wheezing, bronchoconstriction, increased secretion, dispnea, slight chest pain, cough, edema of the lung
Gastrointestinal system	Increased salivation, anorexia, vomiting, abdominal cramps, epigastric and substernal tightness (cardiospasm) with "heartburn" and eructation, diarrhea, tenesmus, involuntary defecation
Sweat glands	Increased sweating
Striated muscles	Easy fatigue, mild weakness, twitching, fasciculations (more pronounced at the side of exposure to the liquid), cramps, generalized weakness including respiratory muscles, dispnea, cyanosis
Central nervous system	Giddiness, tension, anxiety, tremor, restlessness, emotional lability, excessive dreaming, insomnia, nightmares, headache, tremor, apathy, withdrawal and depression, slow wave bursts of elevated voltage in EEG (especially on hyperventilation), drowsiness, concentration difficulty, slow recall, confusion, slurred speech, ataxia, generalized weakness, coma with absence of reflexes, Cheyne-Stokes respiration, convulsions, respiratory and circulatory centers depression, dispnea, cyanosis, fall in blood pressure
Circulatory system	Bradycardia, decreased cardiac output, cardiac arrest, vasomotor center paralysis

to humans is related to its lower toxicity: LD_{50} dermal >4000 mg/kg and the estimated lethal oral dose for a 70 kg man is 60 g, compared to 6.8 to 21.0 mg/kg dermal LD_{50} for parathion and lethal oral dose of 0.1 g. This is not the case if malathion is mixed with maloxon.[30]

As it was mentioned before, compounds containing a P=S nucleus, such as parathion, malathion, etc., must be activated by the metabolic change of S by O, which is performed by mixed function oxidases of the liver and intestinal wall. Such compounds are called indirect inhibitors of cholinesterase activity.

Symptoms of OP intoxication are given in Table 5. Death is due to asphyxia in some instances and to cardiovascular failure in others. In many cases only a few of these symptoms are observed. The interval between exposure and onset of the symptoms may be as short as a few minutes but is usually 1 to 2 h. In cases with predominantly percutaneous intake, this interval is prolonged. Thess et al. report a case of clinically unrecognized, protracted poisoning that started 6 h after the working day of a tractor sprayer ended; he had handled wofatox-concentration 50 and wofatox-spraying solution.[33] Rarely the interval exceeds 24 h.

Other symptoms, such as fever, are also reported, but they are atypical. The respiratory failure results from a combination of respiratory tract blockage from excessive secretion of the salivary glands and respiratory tract, possible bronchoconstriction, and paralysis of the respiration area of the brain stem. According to Zakurdaev the toxic myopathy, e.g., paresis and paralysis of the respiratory muscles, leads to the development of respiratory failure in 5 to 7 d after a severe intoxication.[34]

Organophosphorous compounds generally have been considered little or no cause of primary irritation. Accordingly, relatively few records on the occurrence of skin lesions related to these pesticides are kept. However, the Matsushita et al. survey found that, of all the patients affected by pesticides, the incidence rate of contact dermatitis provoked by OP compounds was 36.5%.[64] From the case analysis, causative OP compounds were found to be DDVP, salithion, simithion (fenitrothion), phosvel, cyanox, kilval, diazinon, and malathion.

The OP effects on the immunological reactions, which may influence morbidity, are very important. Milby and Epstein found that agricultural workers exposed to malathion were sensitized to the intermidiary product diethyl fumarate, and when this compound was decreased in the manufactured product, the incidence of subsequent sensitization also decreased.[65]

Allergic effects due to exposure to OP compounds are described. Nevertheless, Ganelin et al., Davignon et al., and Gardner and Iverson believe that for asthmatic patients such exposure should not be considered an additional risk of increased bronchial sensitivity.[66-68]

Unless exposure causes death, most neurological effects are reversible dependent on ChE inhibitors. Local and less severe effects do not usually last more than one day. Myosis also disappears in less than 1 week, and most other symptoms diminish over the next 6 to 18 d. The symptoms of the intoxication are related to the rate of ChE inhibition and the rate at which the inhibitor itself is destroyed or removed from the tissue. Both these factors are related to the chemical structure of the compounds and eventual presence of byproducts.

Diagnosis of the intoxication can be difficult in mild cases, when only myosis, nausea, vomiting, weakness, headache, and giddiness are observed. In such cases a good anamnesis and ChEA determination will help very much. Unusual toxicological features in poisoning by fat-soluble OPs are reported by Davies et al.[4] In 5 suicidal patients who used dichlorfenthion, the initial symptoms were mild, and cholinergic crisis appeared at 40 to 48 h after ingestion. Two patients died and in the other three ChE inhibition persisted for 48 d. In one patient almost total inhibition of both plasma and erythrocyte ChE was noted for 66 d. This corresponded to pesticide presence in fat tissue at the 54th d and in blood at the 75th d. The levels of dichlorfenthion in adipose tissues in this case were, respectively, 65, 58, and 0.63 ppm on d 4, 7, and 54 after poisoning. Leptophos is more fat soluble and produces a more protracted clinical picture.

In household situations, OP compounds provoke mostly severe acute intoxications. Such intoxications are described by Golden et al., Favre, Gervais et al., Gupta and Patel, Tefik, Bledsoc and Seymour, Cattle.[35-41]

The acute cases are almost always severe, often with lethal issue. The possibilities of a successful treatment increase with the reactivation of cholinesterase, when the patients are duly hospitalized. Gaultier et al. reported a case with an intake of 800 mg parathion.[42] The treatment was successful with methyl-α-pyridinaldoxim, atropin, and reanimation administered successively.

The late diagnosis as well as complications with pneumonia, sepsis, etc. raised serious problems in the treatment.[43] In spite of the intensive therapy, 12.7% of the patients with poisoning by OP compounds die from heart and lung insufficiency.[44]

Tabershaw and Copper investigated the consequences of acute poisonings with OP compounds.[45] A group of 114 subjects poisoned before 3 or more years were investigated; 6 of them had had severe intoxication, 54 moderate poisoning, and 54 light poisoning. Of the group 43 had complaints 6 months after the accident, and in 13 the complaints persisted up to the moment of the examination. Gastrointestinal complaints, eye and brain disturbances, cardiorespiratory, neuropsychic, and other effects were registered. In 8 subjects eye disturbances were found, including longsightedness in 4 of them. Intolerance toward the odor of pesticides manifested in 20 of the subjects with such symptoms as nausea, vomiting, headache, etc. Plasma ChEA was increased.

West published the results of a follow-up study on the sequelae of poisonings with OP pesticides.[46] She found that one fourth of the subjects intoxicated had complaints six months up to several years after the accident. The nervous symptoms were attributed to cerebral anoxia.

B. CHRONIC INTOXICATIONS

Chronic intoxications are rare, because OPs are not highly cumulative. The same symptoms as in acute intoxications are found but are less pronounced: headache, giddiness, insomnia, weakness, increased sweating, nausea, loss of appetite, tremor, and nystagmus.

In epidemiological studies, most frequently reported are liver, renal, skin, cardiovascular, hemopoetic, and respiratory disturbances, as well as aggravation of existing ill health conditions.

The literature data on chronic OP effects are difficult to evaluate because of the interference other toxic substances have during production or other pesticides have during application.

Hartwell and Hayes reported observations on workers in two plants that produce phosdrin, methylparathion, and ethylparathion.[47] The ChEA inhibition has been a frequent finding; however, clinical syndroms appeared very seldom. From 41 cases with inhibited ChEA, 17 have shown different symptoms of intoxication.

Faerman traced the conditions of 179 persons employed in an OP compound production plant: mercaptophos — 50 subjects, chlorophos — 49, methylethylphos — 48 and metaphos — 32.[48] OP concentrations in the air of the working rooms exceeded two to three times the MAC values. A possibility for skin resorbtion existed, as well. The ECG has shown bradycardia and sinus arythmia. According to the author, the ECG and oscillographic studies suggested the presence of an increased vagus tonus. Neurologically, a vegetative dystonia was found, demonstrated by a red dermographism, acrocyanosis, and positive orthostatic reflex, which the author related to the toxic effect of OP compounds. Gastrointestinal tract dyspeptic phenomena were also registered. The functional state of liver showed disturbed proteinsynthesis and hydrocarbon function (elevated blood sugar content, significantly higher α_1 and α_2-globulins together with decreased albumin quantity and a lower albuminoglobulin ratio). At fractional investigation of stomach content, secretory disturbances, such as increased or decreased acidity up to full achylia, were found. Similar changes were reported by the same author some years later.

Kudo investigated 182 agricultural workers engaged in OP application.[49] In 63.9% of them, the pesticide content in blood was 0.004 to 0.520 mg/kg, with a mean of 0.01 mg/kg. Chronic intoxication was diagnosed in 14 workers. LDH, LAP, and aldolase activity in blood serum were significantly below the normal values, especially working with methylparathion. Ophthalmological impairments were observed in two workers. Pyramidal and extrapyramidal damage and cerebral ataxia were also found, as well as deep injuries in the sensory system and parasympaticus (hyperactivity, manifested by myosis and hyperemia of the face).

Women working in rice and barley fields, and exposed to OP pesticides, averaged a 5 to 10 kg decrease in body weight, menstrual disturbances, amenorea, and sterility (mainly transient).[50]

Changes in some biochemical parameters, such as an increased amino acid levels and higher total serum protein have been attributed to the liver function disturbances. There were changes in the albumin globulin index, the serum enzyme activities of aldolase, the alkaline phosphatase, SGOT, SGPT, asparagin amino transferase, and ornitincarbamyl transferase as well.[51-56]

Exposure to OP compounds leads to increased histidine and valin content in the blood and a tendency toward increased cystin, lysin, arginine, alanine, and others.[53]

Decreased reabsorption of phosphorus, due to impaired renal function, has been observed. Urinary amino acid studies demonstrated increased excretion of arginine, ornitine, and lysin.[57-59]

Functional effects on the cardiovascular system such as bradicardia, hypotension, and electrocardiographic changes have been established by other authors.[60,61] Intestinal disturbances such as hypoacidity or achylia and occasional acidity are also reported.

Parathion was supposed to produce aplastic anemia.[62] During chronic exposure leukopenia is reported often, mainly when combined exposure takes place.[63]

Edmundson and Davis found that the active ingredient in naled formulations was the primary sensitizing agent, and its hydrolytic products caused no such reaction.[69]

Ercegovich published a review on pesticide immunological interactions.[70] He concluded that sensitization to high doses of pesticides, as evidenced by dermatitis, may be more prevalent than previously supposed. However, there is no sufficient information to confirm that exposure to low doses of pesticides is actually responsible for hypersensitivity. Several groups produced antisera responsive to protein conjugates of parathion and malathion.

On the basis of the information presently available, OP pesticides are not a serious threat to the immunological defense system of human body. Abnormal exposure to these agents undoubtedly induces hypersensitivity reactions, primarily of the nature of cutaneous manifestations, but there is no evidence that they readily alter the defense mechanism of the body.

C. NEUROTOXIC EFFECTS

The neurotoxic effects from OP insecticide exposure can be classified as either directly related to ChE inhibition or delayed neurotoxic actions.

Most OP poisoning effects are directly related to ChEA inhibition. The mechanism of toxicity involves the inhibition of ChEA at cholinergic nerve synapses with the resulting accumulation of acetylcholine. Prolonged cholinergic stimulation may create a muscle necrotizing effect. Some reported cases with flaccid paralysis and rapid recovery are related to the cholinergic effect but not to neuropathy (malathion, omethoate, parathion, merphos).[1]

This effect is dose dependent. A critical inhibition of AChE activity is by 85%.

Pathological changes are initiated at the motor endplates over a period of 2 h and by the 24th h a generalized breakdown of muscle fiber structure is evident.

D. DELAYED NEUROPATHY

OP-induced delayed neuropathy (OPIDN) is a syndrome, caused by some, but not all, esters. It is characterized by a delayed manifestation of the clinical effects, one to three weeks after the beginning of the intoxication. It is called peripheral distal axonopathy.

The resulting paralysis is caused by concurrent degeneration of the distal regions of long, large diameter axons in the peripheral nerves and spinal cord.[71,72]

Johnson first associated the delayed neurotoxic effects, attributed to the OP insecticides, with the inhibition of the so called neuropathy target esterase (NTE).[11] It is presumed that the process initiates by covalent phosphorylation of the active center of NTE. Aging of the inhibited NTE (Figure 1) is suggested, but still unproven with all the neuropathic agents, to be the second essential step in the inhibition process.[73]

According to Johnson, after the interaction with an OP, the protein called NTE may exist in three forms.[74] The total of these three forms (T-NTE) could be expressed by the following equation:

$$(T\text{-}NTE) = (Ca\text{-}NTE) + (UI\text{-}NTE) + (MI\text{-}NTE)$$

where Ca-NTE is catalytically active, UI-NTE is unmodified inhibited, and MI-NTE is modified inhibited. The modification involves a bond cleavage and the generation of a negative charge. Extensive research concluded that initiation of OPIDP depends on the generation of a substantial amount of MI-NTE.

The substances commonly causing delayed neuropathy in man are triaryl phosphate esters used in hydraulic fluids; they have no AChE activity and they are not pesticides. Table 6 lists the OP pesticides for which reasonable evidence exists that they have caused delayed neuropathy in man.[1]

Predominant motor paralysis affecting the distal muscles of the limbs, minimum sensory abnormalities, and calf pain preceding the onset of weakness are typical for polyneuropathy caused by OP compounds. The electrophysiological findings of partial denervation, with the surviving fibers conducting at normal rates and the pyramidal tract signs noted during the late stages of the illness, are also typical.[75]

In a study by Lotti et al. on workers occupationally exposed to the organophosphorous defoliants IDEF and merphos for several weeks, no abnormalities were found, and there were

TABLE 6
Organophosphorous Pesticides Reported to Cause Delayed Neuropathy in Man[1]

Pesticide	Number of cases	Reference
Mipaphox	2	Bidstrup et al.[77]
Leptophos	8	Xintaras et al.,[83] FAO/WHO,[84] Murphy[78]
Methamidophos	9	Senanayake and Johnson[75]
Trichlorphon	many	Shiraishi et al.,[85] Hierons and Johnson,[86] Johnson[87]
Trichlornat	2	Jedrezejowska et al.,[88] Williams[89]
EPN	3[a]	Xintaras and Burg[90]
Chlorpyriphos	1	Lotti and Moretto[91]

[a] Moderate effects only and possible other etiological factors.

no consistent changes with time in the electrophysiologic findings in the subjects.[76] In all workers, the maximum motor conduction velocity and terminal latency in the right ulnar and peroneal nerves were normal, as were the amplitude and latency to peak of the right ulnar and sural action potentials. There were no significant differences in the values obtained in the examinations before and after exposure period. No evidence of impaired neuromuscular transmission was demonstrated. Needle electromyography responses showed no abnormalities. The authors used the measure of NTE inhibition in lymphocytes as a monitor of the occupational exposure to DEF and merphos. NTE was inhibited about 40 to 60%, approaching the pre-exposure values 3 weeks after the end of the exposure. Since no effect on the physiology of the peripheral nervous system was observed, the authors concluded that equally high levels of inhibition of NTE activity (70 to 80%) are required in humans, as in animals, to trigger the neurotoxic response.

Neurotoxic effect is independent on the inhibition of ACh.[11] Axonal degeneration is followed by degeneration of myelin sheath cells in the peripheral nerves, and in some cases, the degeneration of tracts within the spinal cord.[72]

Bidstrup et al. first reported cases of delayed neurotoxicity from an organophosphorous pesticide in 1953.[77] Two research chemists, working with the experimental pesticide mipaphox, experienced weakness and unsteadiness of gait two or three weeks after recovery from an episode of acute poisoning. One developed bilateral foot drop from which he recovered, while the other progressed to a flaccid paralysis of the lower extremities, and had not improved after two years.

Leptophos (phosvel), which is known to cause delayed neurotoxicity, has also produced paralysis in humans and buffaloes.[78]

Neuropathy, caused by EPN and leptophos is related to repeated occupational exposure with inadequate precautions; apparently slight cholinergic effects were often experienced. A few cases with methamidophos and trichlorfon involved substantial occupational exposure. It caused severe acute poisoning prior to neuropathy development, but most of the cases involved accidental or deliberate ingestion of quantities that might well have been fatal but for medical intervention.

In case of suicide by a mixture of pesticides including chlorpyriphos (about 20 g diluted in petroleum distillates), Osterloch et al. found that erythrocyte and peripheral nerve AChE levels were about 22% of normal and that nerve NTE was about 30% of normal.[79] They predicted that treated survivors of severe poisoning by chlorpyriphos might develop delayed neuropathy. This was confirmed in another case of suicidal poisoning with chlorpyriphos (about 300 mg/kg/b.w.). Very low levels of lymphocyte NTE were found 30 d after this heavy intoxication. Typical moderate polyneuropathy developed in the following days (Lotty and Moretto, citation by EHC[1]).

De Jager et al. described a patient with a purely motor neuropathy, with muscular wasting and weakness of both upper and lower limbs, after ingesting an extremely high quantity of parathion (150 g), dissolved in methylalcohol.[81] EMG demonstrated signs of acute denervation with normal conduction velocities in motor and sensory nerve fibers. This indicated predominantly axonal degeneration. Changes in sural nerve biopsy consisted of slight axonal degeneration, especially of the larger myelinated fibers, and some segmental demyelination. The authors are convinced that this patient had a delayed polyneuropathy. They suggest that only a massive exposure to parathion with extensive artificial respiration and charcoal perfusion, as it was in the cited case, might cause polyneuropathy; in experimental intoxications caused by ACh excessiveness, animals die from cholinergic symptoms long before neuropathy can develop.

Polyneuritis and paralysis have been reported as a consequence of intoxication by malathion, parathion, merphos, and omethoate.[92-96] In these cases, interference of impurities or other etiological factors is possible. According to Lotti et al. omethoate has negligible potential to cause delayed neuropathy.[97]

Hyporeflexia was observed in agricultural workers exposed to OP compounds. It is considered a possible sensitive test for diagnosing chronic intoxications with OP compounds.[98]

Preliminary results of the International Epidemiological Study on Health Effects of OP Pesticides demonstrated some changes, indicating toxic neuropathy in exposed agricultural workers.[99,100]

Short reviews on OP neurotoxicity were prepared by Kaloyanova, Batora, Seppalainen, Johnson.[100-103]

E. ELECTROMYOGRAPHIC STUDIES

According to Roberts and Wilson and Jager et al., subclinical neuropathy is demonstrated by EMG changes.[104,105] Neuropathy target esterase of human lymphocytes was proposed as a predictive monitor for delayed neuropathy effects.[106] The importance of EMG for early diagnosis has been confirmed by other research, conducted by Roberts.[107] Organophosphorous pesticide factory workers have been examined electromyographically, and a relation between work with OP compounds and low voltage EMG in response to supramaximum ulnar nerve stimulation has been demonstrated. Workers with low voltage EMG also average low conduction velocities in both the fastest and the lowest motor nerve fibres. A comparison of the maximum conduction velocity of motor nerve fibers in control and organophosphorous exposed workers showed that the latter group averages 10% lower velocities than the control group.[107] Drenth et al. found in approximately 40% of 102 male agricultural workers abnormal but not persistent EMG patterns and no ChE inhibition.[108]

Six men occupationally exposed to OP pesticides were examined electromyographically over a period of 7 to 9 months. Despite the absence of clinical signs and symptoms of cholinesterase effects, the EMG voltage varied within a pattern, reflecting exposure. The EMG results were used to illustrate the way individual exposure of workers can be monitored and reduced by improving precaution measures.[109]

To investigate reported nervous effects on chronic exposure to organophosphorous pesticides, 25 µg/kg mevinphos daily was administered to male subjects for 28 d; results were compared to those of a control group. At the end of the exposure, a 7% decrease in slow fiber motor nerve conduction velocity and a 38% increase in Achilles tendon reflex force were found. No effect on neuromuscular transmission was observed. Red blood cell cholinesterase depression was 19%. These results confirmed similar deviations, found in pesticide workers.[110] Jusić et al. evaluated EMG neuromuscular synapse testing and neurological examination for early detection of organophosphorous pesticide intoxication.[111] Two groups of healthy agricultural workers and one group of healthy spraymen have been exposed to various organophosphorous pesticides of different intensities. No significant difference between the exposed group and the control groups was found. The electrically evoked muscle potential series in exposed workers remained as constant as those recorded at the control group. The frequency of different type of amplitude changes was the same in exposed and in control groups. Neurological records showed no significant deviations in the exposed workers. Only the subject with low cholinesterase activity demonstrated inconstancy.

Further studies in this field are evidently necessary.

Hussain et al. found that both EMG and AChE determinations are suitable for monitoring occupational exposure to anticholinesterase agents. In view of the potential advantages of EMG, its use for this purpose should be further developed and improved.[112]

F. CENTRAL NERVOUS SYSTEM

OP pesticides affect some functions of the CNS. Cholinergic and noncholinergic mechanisms are involved in a variety of cases and long-term damages, including pathomorphological phenomena. Karczmar published a comprehensive review on this matter.[113]

Initial reports of persistent CNS manifestations included impaired memory, depression, impaired mental concentration, schizophrenic reaction, and instability lasting from 6 to 12 months. The 16 subjects had been exposed to organophosphorous insecticides for 18 months to 10 years.[114]

Later on, Dille and Smith described two cases with mental disturbances.[115] The study concerned pilots working mostly with OP. One of them developed depressive phobia, and the other — fear and emotional instability. As a sequel, schizophrenic reactions have been demonstrated after heavy acute intoxication.[116] Kovarik and Sercl reported neurastenic manifestations in survivors after acute intoxication as did West.[117,118]

Extensive studies performed by Metcalf and Holmes found no difference between exposed and control subjects in terms of such hard neurological signs as sensory or motor deficits.[61] Exposed men showed more of the so called "soft" neurological signs, such as motor coordination deficiency and oculomotor imbalance. The psychological test battery consisted of the Wechsler Adult Intelligence Scale (WAIS), the Benton Visual Retention Test, and the Story Recall Task. The results indicated that the disfunctions most clearly seen in the exposed group include a disturbed memory and a difficulty in maintaining alertness and focusing attention. The exposed subjects mostly use a variety of such compensations as delay, avoidance, inappropriate giving up, and slowing down.

Interviews independently check up on some of the psychological testing results and uniquely gather special information. The aim is to obtain a systematic and relatively objective view of the exposed men, their feelings about themselves and their symptoms (if any), any changes over time of which they may be aware, and their attention to industrial health practice, for example. Subjects with histories of multiple or severe exposure complain directly and give evidence of being slowed down and less energetic and having increasing memory difficulties. They also have slowness in tapping and calculation and greater irritatability than the minimum exposed group.

Maizlish et al. investigated the neurobehavioral effects of short-term, low-level diazinon exposure among 99 granule applicators.[119] They were tested before and after their work shift with a computer assisted neurobehavioral test battery. The post-shift median diazinon metabolite diethylthiophosphate (DETP) level for the exposed and control subjects was 24 and 3 ppm, respectively. The whole body exposure was calculated to be 2.1 and 0.03 mg, respectively, with a mean duration of diazinon application 39 d (SD = 12 d) before testing. No adverse DETP-related changes in pre- or post-shift neurobehavioral function were found, although Symbol-Digit pairing speed was slower among the applicators as a group. The prevalence of 18 symptoms, possibly related to diazinon exposure, was not elevated among applicators.

G. BEHAVIOR CHANGES

Behavioral studies have been conducted to assess the psychiatric manifestations in workers with low occupational exposure to organophosphate compounds and no obvious signs of toxicity. Commercial pesticide sprayers and farmers recently exposed to organophosphate agents were compared to control subjects on personality tests, a structures interview, and cholinesterase levels. Depressive symptoms were assessed by the Beck Depression Inventory. Anxiety was measured by the Taylor Manifest Anxiety Scale. The commercial sprayers, not the exposed farmers, showed elevated levels of anxiety and lower plasma cholinesterase levels. Following are the principal behavioral changes: (1) difficulty in concentration, (2) slowed in information processing and psychomotor speed, (3) memory deficit, (4) linguistic disturbance, (5) depression, and (6) anxiety and irritability.[120]

On the basis of a comprehensive literature review, Maizlich concludes that persistent neuropsychiatric sequelae occur in approximately 4 to 9% of those with acute OP pesticide intoxication.[121] Case reports suggest that symptoms generally resolve within 1 year after acute poisoning, although subtle functional changes were present 9 years later in at least 1 study. A limited number of cross-sectional epidemiological studies on workers with long-term employment or previous poisoning reveal subtle neurobehavioral effects, such as poorer performance on a battery of subtests involving intellectual functioning, academic skills, alertness maintenance, visual intelligence, and anxiety. However, these studies were not controlled for level of exposure, age, education, and alcohol consumption. Reports on asymptomatic workers with slight but measurable ChEA depression present contradictory results according to Maizlich.

Behavioral studies not only reveal the real danger of some agricultural occupations from health point of view, they also reveal difficulties in work performed, for example plane crashes in cases of aerial application.

The possible relationship between aircraft incidents and pesticides effect is discussed by Reich and Berner.[122] They found, from 12 accidents with pilots, 8 cases with ChEA inhibition. Wood et al. reported an accident in which disturbances in coordination and the ability to regulate the speed of the airplane were found in pilots.[123] Heat stress in hot climate conditions seems to play a very important role in aircraft accidents.[124]

Durham et al. investigated 53 subjects with differing degrees of exposure to OP compounds.[125] Out of several vigilance tests, only 1 could establish some changes. No disturbance in the reaction time was detected; only in 2 subjects, hospitalized for OP pesticide poisoning, were disturbances in vigilance observed — but always accompanied with other symptoms.

In the clinical cases of poisonings by OP pesticides, the behavioral impairments have been characterized by the following symptoms: fatigue, irritability, coordination difficulties, and slow thinking processes. Often these symptoms persist more than 1 year after the exposure.[126] Kaloyanova et al. found out a slight deviation in reaction time and memory potentials in some behavioral tests on agricultural workers exposed to OPs.[127]

H. EEG STUDIES

EEG investigations have defined characteristics of CNS disturbances in OP exposed workers.[61,128,129] They found no increase in the incidence of hard EEG abnormalities such as spike activity or local slowing. The outstanding finding in EEG studies was a high incidence of low- to medium-voltage slow activity in the theta range, that is, 4 to 6 Hz activity occurred during light drowsiness in brief episodes of 2 to 4 s duration. Visual and auditory evoked responses have been examined to test the hypothesis that OP exposure might result in disturbances of CNS information processing capability. Sensory evoked responses provide a large amount of information and can be accomplished quickly, with a large number of people. Auditory evoked responses show more variability than the usual responses. They both have shown trends toward lower amplitudes and longer peak latencies in the exposure group.

Recently developed methods in computer analysis would yield specific information and permit the rapid identification of changes. They could possibly use EEG as a sensitive early index of CNS impairment.[129]

Increased delta activity, increased delta and theta slowing, decreased alpha activity, and increased rapid eye movement during the sleep of the exposed population are persistent and parallel to behavior changes. While it appears that OP anti-ChEs may induce long-lasting (up to one year) EEG changes, the currently available data relatively leave open the question of whether or not OP drug exposure evokes delayed psychic effects.[113]

The specific EEG findings, sensory evoked response disturbances, sleep disturbances, and deep midbrain effects of OP-anticholinesterase compounds are of major importance in the production of CNS changes. Whether long-term exposure to OP compounds can induce irreversible or only slowly reversible brain disfunction needs further intensive study.

I. EYE AND VISION

Eye and vision impairment in acute intoxication are well known. Long-term exposure to OP seems to produce visual impairment and eye abnormalities as well. Plestina and Piukovic-Plestina reviewed existing publications; they summarized most of the available literature data about eye and vision impairments attributed to anticholinesterases or pesticides in general.[130] They also gave a brief summary of their own investigation. Visual field stenosis (narrowing visual field), progressive myopia, astigmatism, edema and atrophia of the optic nerve, lenticular changes, cataracta, impaired balance, slower pupillary sphincter reaction, and low dark adaptability have been attributed to OP. Research revealed mild constriction of the peripheral visual fields and a somewhat lower dark adaptability in exposed workers, which might be connected with a possibly slower reaction of the pupillary sphincter.

Abundant experience with powerful anticholinesterase miotics gathered by clinicians seems to add little support to the idea that anticholinesterases produce other eye impairments, except transient lenticular changes.

At present the authors feel it is prudent to accept the findings that long-lasting, mostly reversible eye and vision impairment might be encountered in highly exposed subjects. Most of these changes could be explained satisfactorily through the known cholinergic mechanism, while others may be completely independent of the cholinesterase inhibition. The findings for workers heavily exposed to organophosphorous insecticides for a very long time are not consistent with the suggestion that this group of chemicals might be a cause of severe eye troubles in persons whose only exposure is environmental.

Revsin suggests that all OPs penetrating the blood brain barrier will disturb visual function by selectively blocking the directional sensitivity of the visual integrative neurons in the thalamus.[131]

J. LONG-TERM EFFECTS

There are no epidemiological data in relation to OP carcinogenicity.[132] Only animal studies give some indication of potential long-term effects for man.[99] Kaloyanova published a review of the experimental works on long-term effects.[133]

K. EFFECT OF INDIVIDUAL COMPOUNDS ON HUMANS

Parathion is one of the most hazardous representatives of the OP group. It is highly toxic and often causes lethal poisonings, mainly because the symptoms of intoxication are manifested after a latent period.

Even a very short exposure to parathion, if massive, can provoke severe intoxication. Such was the case reported by Thiodet et al. of a 17-year-old worker, who after 3 d work with parathion, manifested the following symptoms of poisoning: severe headache, loss of consciousness, lung edema and azotemia and glycemia.[134] Carini et al. reported the parathion poisoning of an agricultural worker with a liver injured from alcohol abuse; only a 4-h exposure to a low concentration of parathion caused a poisoning with lethal issue.[135]

The data of Orlando et al. show that at continuing contact, an adaptation toward parathion occurs.[58] Studies with 537 agricultural workers showed that 40% of the subjects exposed to parathion had a progressive decrease in the symptoms manifested at the first contact. Most probably, this is only a seeming adaptation, since changes in the clinical laboratory investigations are observed. In subjects with moderate intoxication, a deficit in the resorption of phosphates was found on the 18th day, attributed by the authors to reversible tubular injuries.

After an agricultural worker was exposed to parathion for 3 years, such symptoms as dizziness, severe headache, vomiting, diarrhea, and sweating were observed. A half-day occupational exposure, once in 15 d during the course of 4 years, resulted in anorexia, diarrhoea, many days of epigastralgia, and breathing difficulty.[136]

For the poisoning diagnosis and exposure evaluation, paranitrophenol in urine and ChEA was determined.

The results of blood and urine analysis showed that parathion concentration in blood correlates well with paranitrophenol concentrations in urine ($r = 0.56$). At the same time, no correlation between the levels of parathion in serum and cholinesterase, or of cholinesterase and excretion of paranitrophenol in urine, was found.[137,138]

Milthers et al. analyzed the symptoms of severe acute parathion intoxication in 20 patients of the Center for Control of Poisonings in Copenhagen from 1952—1962.[139] The following symptoms were found: gastrointestinal — in all patients; neurological (cramps, coma etc.) and vertigo — in 17 patients, kidney — in 15 patients, and respiratory — in 12 patients. The latent time for the symptoms to appear was from 20 min to 24 h. The death of 8 of the patients was caused by: shock — 2 patients, lung edema — 1, pneumonia — 1, lung artery embolia — 1, cramps — 2, and atelectase with pancreatic necrosis — 1 patient.

Sometimes, besides the known symptoms of OP compound intoxication, some rare symptoms are observed. They are ascribed to the effects of parathion: death due to thrombosis of vena saphena interna without any intravenous injection, being applied (Frank, cit. after Kaloyanova-Simeonova et Fournier), hypothermia up to 34°C combined with deep hypotonic coma, and hyperthermia with pupil dilatation.[140-142] Monov describes the picture of apoplexy with hypertension and anisocoria in a woman poisoned by parathion.[116]

Wakatsuki reported paralysis of the larynx, due to parathion intoxication.[143] Ishikawa observed an atrophy of eye nerve following parathion.[144]

A rather dangerous formulation is methylmercaptophos(sistox). In view of the great number of intoxications it caused, it was forbidden in the U.S.S.R.

Rosin analyzed 222 case histories of persons intoxicated by methylmercaptophos and established that the dermal route of penetration was the most probable compared to inhalatory or oral penetration, it had a longer latent period (3 to 8 h).[145] The earliest symptoms of intoxication are giddiness with blurred vision, a headache located mainly in the parietal-temporal area, and general weakness. Later on nausea, vomiting, pains in the epigastric and abdominal area, and diarrhea appear. Pallor of the skin and membranes and furred tongue are observed, as well as moderate tachycardia and less often bradycardia, dull cardiac sounds. Single crepitations are detected in the lungs. Better expressed are the changes in the nervous system — apathy and sleepiness or on the contrary, excitement and irritativeness. Almost all sufferers show a moderate sweating, cooling extremities, and cyanosis of fingers and toes. In more severe intoxications, these symptoms are better expressed.

Not infrequently, in severe forms of intoxication, loss of consciousness, weakness, vomiting (sometimes irresistible), and profuse diarrhea are noted. Cramplike twitches of the facial muscles, fingers, and toes and clonic-tonic cramps of the upper and lower extremities are characteristic. Pupil reaction is absent in the majority of the intoxicated, the tongue is furred, and the skin is pale and cyanotic. Rosin considers the absence of myosis, salivation, bronchial spasm, and bronchorrea manifestation, which are leading in the symptomatology of OP compound intoxications. Of 121 workers exposed to metasistox for 1 to 18 d, gastrointestinal manifestations were found in 70, fatigue and weakness in 29, and headaches in 15; hypersecretion, tremor, and ataxia were observed in 5 subjects. The inhibited ChEA serum recovered within 30 d after treatment.[146]

Tilsner described metasistox poisonings with prolonged development.[147] The first symptoms appeared 60 h after the intoxication, and death after 81 h. Fatty degeneration of the liver was found, which could be a possible cause of the death.

One case, described by Redhead, is quite characteristic of occupational poisonings in agriculture.[148] It concerns a worker, in a specialized group, engaged in plant spraying over the course of 5 years; his daily exposure was from 20 min to 6 h. During the last 6 weeks of this period he worked as an adviser to aerial metasistox spraying. At the same time he was engaged with the preparation of the solution and the cleaning of the containers. He felt headaches, nausea, and dizziness 2 weeks after the beginning of this work; these complaints grew from day to day up to the weekend. They renewed with the beginning of the next work week, when new symptoms appeared, such as anorexia, loss of concentration, and lower cholinesterase activity — 7 units/ 100 cm^3 vs. the standard 40 to 80 units/100 cm^3. The ChEA was up to 43 units 8 d after discontinuing this work, and 13 d later the person recovered.

One worker has shown acute erhythrodermia, followed by polyneuritis, two weeks after stopping a continuous metasistox spraying[149]

High temperature (36 to 40°C) potentiates the toxic effects of anthio upon cotton farming workers.[150]

An investigation on 12 agricultural workers, occupationally exposed to malathion during 6 months, revealed ChEA inhibition and SGOT, SGTP, and serum aldolase depression in 11.[151]

Gardner and Iverson investigated 119 subjects who had been exposed to malathion to a different degree.[68] After aerial application of 95% solution against mosquitoes during an encephalitis epidemic in Texas in 1966, 6 of the subjects showed signs of nausea, headache, and weakness, without ChEA inhibition.

Selassie and Lester described 8 cases of malathion intoxication among workers unpacking the preparation; they became dusty when the wind blew against their faces.[152] The patients were transported to a hospital 130 km from the farm. Two of them died during the transportation, and the others were hospitalized 8 h after the exposure.

Four acute malathion poisonings in children were reported in Israel in 1969. They had washed their hair with malathion solution. The laboratory analysis revealed hyperglycemia and glucosuria, which rapidly recovered after atropine and pralidoxine treatment.[153]

Stevens reported inhibition of liver microsomal metabolism due to malathion exposure.[154]

To assess the effects attributed to malathion that escaped from an overheated tank at a chemical plant, Markowitz et al. surveyed sailors presumably exposed on board a nearby tanker, as well as control group of sailors.[155] The exposed subjects more commonly reported five body systems or functions as affected, compared with the controls. The head, the eyes or vision, and bowel movements are known to be affected in OP intoxications. Verification of OP intoxication was reportedly based on substantial depression of plasma and red blood cell ChE. Baker et al. demonstrated that ChE depression in field personnel exposed to malathion was greatest in those using the pesticide brands with the highest concentrations of isomalathion and other degradation products.[30]

Pesticides meant for home use raises the question of hazard, not only are the people performing this operation at risk, so are the inhabitants. Vandekar reported his observations during folithion application for mosquito control in Nigeria.[156] He established the existence of a real danger for intoxication of children living in such houses after treatment. The development of the intoxication is favorable, leading to a rapid recovery. A relationship was noted between the clinical symptoms and ChE inhibition.

Menz et al. investigated workers who produce dichlorphos at an approximate concentration of 0.7 mg/mg^3 in the air of the working environment.[157] Observations were performed in the course of 1 year. No deviations were found either in the hematological and biochemical indices or in the urine analysis, except for a moderate ChE inhibition in plasma and erythrocyte cells. The significance of ChE determination is underlined not only as an indicator of exposure, but as a more sensitive index than the direct determination of DDVP in blood.

Matsushima described occupational dermatitis as an effect of exposure to dichlorphos.[158]

Kandelaki reported 11 cases of intoxication with chlorphos (trichlorfon, dipterex) and trichlormetaphos.[159] The first symptoms of intoxication were headache, dizziness, vomiting, sounds in the ears, general weakness, dull epigastric pains, sleep disturbances, liver pains, pollakiuria, lacrimation, leukocytosis, lymphopenia, and accelerated SRE. After treatment all the patients were discharged from hospital in good condition. After repeated contact with the chemicals, some of the subjects complained of joint pains, headaches, insomnia, general weakness, and stinging in the eyes. Six intoxications by the dermal route were described. At prolonged contact with chlorphos and trichlormetaphos the skin grows coarse, dry, and cracked. Folicular ceratitis, infiltrates, hyperhydrosis, and hyperemia are observed.

Tsapko found functional liver changes at poisoning with chlorphos.[160]

Ikuta described cases of acute poisoning with trichlorphon.[161] A farmer ingested about 65 cm^3 of 50% trichlorphon solution, and the intoxication developed with coma and myosis. Heart hypertrophy was established; no deviations were found in the liver function. The treatment included atropine and pralidoxim. After 38 d polyneuropathy was diagnosed; serum ChEA recovered on the 29th day.

After an acute intoxication with trichlorphon, polyneuritis with shoulder pains, paresis and paralysis of the lower extremities, reduced tactile sensibility, absence of peritoneal and some other reflexes, diffuse muscle dystrophy, and cyanosis were observed.

Babkina[162] and Shutov and Barankina[163] considered the consequences of chlorphos intoxication (dipterex), observed in 3 cases. Paresthesia with leg and arm pains appeared 10 d after severe intoxication. Polyneuritis was diagnosed 2 months later, based on paraesthesia and the presence of motor, sensory vegetative, and trophic disturbances. A slow improvement of the impairments was noted after 2 months.

Holmes et al. reported 2 cases of moderate acute intoxication with meviphos (phosdrin).[164] The symptomatology was similar to that of parathion or zarine intoxication. A rapid excretion of dimethylphos was established; it ended 50 h after exposure to mevinphos. In 1 of the cases, fibrinolysis was registered, and in the other, a marked tendency toward blood hypercoagulation was found, which was underlined by the curve of thrombin production. Hematuria that persisted more than 8 d was observed in 1 of the patients. In both cases the therapy had a good effect.

In the case of severe mevinphos intoxication of an 18-year-old girl, Stoeckel and Meinecke reported plasma ChEA inhibition in the course of 3 months and of erythrocyte ChEA inhibition in the course of 5 months.[165]

Chezzo et al. reported a higher sensitivity to OP intoxications in subjects who had been exposed to OP compounds for many years.[166] The authors explained this phenomenon as sensitization to acetylcholine. ChE activity in such cases was not different from that of subjects who had not been exposed before the intoxication.

The sensitization to phosdrin was confirmed by Bell et al.[167] It concerned a worker, who had twice worked with phosdrin in a vineyard.

Weiner reported a case of bronchial asthma, followed by polysensitization in the course of three weeks, after acute phosdrin intoxication.[168] A short contact with household dust and different chemical substances was able to provoke an allergic reaction.

Perron and Johnson presented data about a case of intoxication, with strong ChEA inhibition, after a 3-week exposure to bidrin.[169] The development of the disease revealed an interesting picture. A respiratory paralysis developed 6 d after hospitalization, in spite of the applied treatment with atropine and pralidoxim. The symptoms required a 22 d hospital treatment to disappear; ChEA reactivation needed several months in addition.

Conyers and Goldsmith described a symptomatic psychosis in a worker who had performed desinsection of sheep with diazinon.[170]

Intoxication with diazinon emulsion was observed in a cattle herd and 3 herdsmen. One of these workers and 32 animals died.[171]

Banerjec[172] described pericarditis resulting from an acute diazinon intoxication; acute OPC intoxication was also reported by Dochev and Latifyan[173] and Constock et al.[174]

L. BIOLOGICAL MONITORING OF OP EXPOSURE

Data on the relationship between exposure and ChE inhibition are shown in Tables 2 and 3. In principle, ChE activity correlates with the amount of pesticides absorbed in the organism. This is the reason ChEA is a specific test for exposure to OP compounds. ChEA inhibition is commonly found in workers exposed to organophosphorous compounds.

Simpson et al. investigated agricultural workers exposed to OP compounds and OCP through processing cotton in Australia.[175] The OP compounds used were methylparathion, mevinphos, and methyldemeton. The chemicals were applied by spraying. In February and March 1972 the whole blood ChEA was determined. In 50% of the subjects it was lower than the normal values, and in 4 workers it was significantly lower. At the repeated determination of ChEA in March, ChEA inhibition was found in 24 of the 30 investigated workers.

Holmes published the results of a survey on Colorado farmers who had been exposed to OP pesticides for a period of 7 years (1955–1961).[176] From a total of 419 investigated subjects, 70 had been hospitalized. ChEA inhibition was established in 40% of the workers. Among the personnel servicing the aerial spraying, this proportion was even higher: in 86% of the mixturers and 90% of the fillers, ChEA inhibition was observed.

Kalic-Philipovich et al. found lower blood ChEA in 17.72% of the workers engaged in the production of OP compounds.[177]

OP metabolites in urine are also very sensitive indicators and in principle, should correlate with the exposure. As yet, the correlation of alkylphosphates with health effects has not been well studied. More studies in this aspect have been done with the parathion metabolite *p*-nitrophenol and, recently, in connection with urinary alkylphosphates.

An occupational study was conducted by Hayes et al. for a firm employing 22 pest control operators (PCOs), exposed to 3 organophosphorous insecticides.[178] The exposure levels measured were less than 131.0 $\mu g/m^3$ for vaponite, 41.0 $\mu g/m^3$ for diazinon, and 27.6 $\mu g/m^3$ for dursban. Hayes et al. analyzed 24-h urine samples for alkylphosphates and showed the presence of metabolites of these 3 pesticides. The effect of this exposure was reflected in a statistically significant inhibition of plasma acetylcholinesterase (AChE) among the PCOs as compared to a sex and age matched control group. There were no significant differences in the mean RBC AChE values of either group.

The authors stated that only the combination of air sampling, urinary alkylphosphate ChE determination, and physical examination presents a balanced picture of the degree of exposure.

The airborn concentrations of guthion 50 WP (azinphos-methyl) were measured near workers spraying with orchard air blast equipment; they were reported to be in the range of 0.02 to 0.11 mg/m^3. The level of guthion on the body patches ranged from 1.8 to 18.8 ng/cm^2. The cholinesterase assays revealed no depression of either RBC or serum ChEA on exposure days greater than 15% of the pre-exposure levels.[179] These did not exceed the variations observed in the control group. No orchardists were found to be suffering from any ailment. The pre-exposure values of urinary metabolites ranged from 42 to 137 $\mu g/24$ h, whereas the spray day 1 values ranged from 117 to 1754 $\mu g/24$ h.

TABLE 7
Clinical Symptoms of Different Grades of OP Poisoning and Corresponding ChEA Values[9,190,191]

Level of poisoning	Clinical symptoms
Mild 60% reduction of ChEA	Weakness, headache, dizziness, diminished vision, salivation, lacrimation, nausea, vomiting, lack of appetite, stomach-ache, restlessness, myosis, moderate bronchial spasm; convalescence in 1 d
Moderate 60–90% reduction of ChEA	Abruptly expressed general weakness, headache, visual disturbance, excess salivation, sweating, vomiting, diarrhea, bradycardia, hypertonia, stomach-ache, twitching of facial muscles, tremor of hand, head, and other body parts, increasing excitement, disturbed gait, and feeling of fear, myosis nystagmus, chest pain, difficult respiration, cyanosis of the mucous membrane, chest crepitation; convalescence in 1–2 weeks
Severe 90–100% reduction of ChEA	Abrupt tremor, generalized convulsions, psychic disturbances, intensive cyanosis of the mucous membrane, edema of the lung, coma; death from respiratory cardiac failure

Excretory levels of parathion metabolites — paranitrophenol correlate with exposure. The peak reached 8 h after the beginning of the exposure. The mean maximum values of *p*-nitrophenol are reported to be 79.2 µg/h with a range of 30 to 168 µg/h. The excretion of *p*-nitrophenol falls to a nondetectable level by the 5th to 8th day after exposure. The increase of ambient temperature stimulates the excretion. The mean amount of parathion absorbed and excreted as paranitrophenol is 2.78 mg for 4 to 8 d. This dose represents 0.02% of the lethal dose (LD$_{50}$) for rats.[180]

As stated by Wolfe et al., the mean parathion respiratory exposure is 0.02 mg/h.[180] Durham et al. studied the correlation between dermal and inhalatory exposure and excretion of parathion.[181] By calculating the excreted *p*-nitrophenol, they found that dermal exposure represented 80 to 90% of the total quantity absorbed in the body. Thus, total urinary excretion of *p*-nitrophenol varied from 0.67 to 7.9 mg. Respiratory exposure was responsible for 0.13 to 0.76 mg, and dermal absorption from 0.54 to 7.15 mg *p*-nitrophenol.

This calculation has been supported by a special volunteer study that eliminates one of the two routes of exposure, the dermal or the inhalatory in different individuals. When skin was well protected, the total *p*-nitrophenol excretion was 0.006 to 0.088 mg, which represented the inhalatory exposure; when the skin was not protected it was 0.597 to 0.666 mg. The estimated dermal exposure represented 0.40 to 1.95% of the exposure potential measured by dermal pads.

For other OP pesticides, limited information is available about specific metabolites. A daily oral dose of 16 mg azinphos methyl for 30 d increased the excretion of anthranolic acid.

Using EMG as measure of exposure is not recommended. As an early indication of neuropathy, the changes in EMG will appear in some period after exposure. Fifty-five cases of urinary schistosomiases in children were treated with metrifonate in 3 doses reaching up to 10 mg/kg/b.w. at 2-week intervals. Repetitive activity was recorded over the thenar muscles in 3 cases, following supramaximal stimulation of the median nerve in the wrist at maximum erythrocyte ChE inhibition.[182] No change in evoked muscle action potential amplitude was detected.

1. Exposure Tests as an Early Indication of Health Impairments

As previously mentioned, OPs act mainly by inhibiting the enzymes cholinesterase and pseudocholinesterase, which are responsible for hydrolyzing ACh at synaptic sites.[183-191] The enzyme inhibition is almost irreversible. The correlation between ChEA and the clinic signs of acute poisoning is reasonable (Table 7).

Signs and symptoms of poisoning by organophosphorous compounds occur when more than 50% of the ChE or erythrocyte AChE is inhibited. The recovery of blood cholinesterase takes about 3 weeks in patients with mild poisoning. However, the recovery of synaptic AChE appears to be very rapid; signs and symptoms disappear within 24 h in patients with mild or moderately severe poisoning. The plasma ChE decreases, but it is normalized more quickly than that of the cell. After a severe intoxication, the reduction of enzymes lasts up to 30 d in plasma and up to 3 months in erythrocytes.[80]

If the rate of ChE inhibition is rapid, the correlation between the inhibition of blood ChE and the severity of symptoms tends to be good. When the rate of ChE inhibition is slow, the correlation can be low or nonexistent. This can happen during long-term occupational exposure, because the body adapts to the high levels of accumulated acetylcholine.

In general, the acute cholinergic effects of severe organophosphorous poisoning correlate well with cholinesterase inhibition. Chronic moderate exposure results in cumulative inhibition of the RBC and plasma enzymes. The appearance of symptoms depends more on the rate of fall in ChEA, than on the absolute level of the activity reached. Workers may exhibit 70 to 80% inhibition of RBC and plasma cholinesterase enzymes after several weeks of moderate exposure, without manifesting cholinergic symptoms. At first exposure, individuals may develop symptoms even after inhibition of cholinesterase activity is less than 30%.

This is the reason why some authors also failed to find a correlation between symptoms and ChE levels.[183-188] ChE screening to predict the health effects of chronic low-level exposure is limited.

Petkova, on the basis of her results in several groups of workers (n = 155) occupationally exposed to OP pesticides during application (tractor sprayers, mechanics repairing spray equipment, sprayers in public health sanitation, and greenhouse workers), stated that ChEA determination does not seem to have a leading role in the assessment of health effects of a chronic exposure.[189] During this study, the depression of serum ChEA, if any, was within the range of physiological variations. A reverse correlation was established between erythrocyte ChE and paranitrophenol in urine after exposure to methylparathion. No correlation with serum ChE was established. Erythrocyte ChEA was accepted as a better biological indicator of chronic exposure.

Roan et al. provide data for the aerial application of OP pesticides.[188] The data suggest that erythrocyte and/or plasma ChE values are not as sensitive indices of ethyl or methyl parathion absorption as the concentrations of these compounds in the serum are. There has also been an indication that assays of serum samples via GLC are even more sensitive than urine *p*-nitrophenol measurements.

Levels of metabolite considered alone are not guides to hazard. Pesticides that have very different toxicities may yield identical acidic metabolites. Thus, the level of metabolite in urine, after it is exposed to sufficient amounts of the very toxic parathion-methyl to depress blood ChEA up to 50%, will be much less than the identical metabolites, to fenitrothion, a compound of about 40 times lower toxicity.[1]

Davies et al. underline that alkylphosphates in urine are very sensitive indicators for OP, but not for individual compounds.[192]

Urinary phenolic metabolites, when present as components of these molecules, are sensitive indicators for exposure to individual compounds.

Exposure to and absorption of guthion 50 WP (azinphos-methyl) were estimated in orchardists who were involved in mixing, loading, and application.[179] Air monitoring and patch techniques were used to estimate exposure; the alkyl phosphate excretion and cholinesterase inhibition were measured to estimate absorption.

All workers had quantifiable levels of alkyl phosphate following exposure, and the 24-h urine samples provided a more reliable estimate than first morning voids. There was no depression of either RBC or serum cholinesterase exceeding the variations observed in the control group. A high correlation was observed between 48-h alkyl phosphate excretion and the amount of active ingredient sprayed.

In the case of parathion exposure, concentrations of dimethyl phosphate in the first urine exceeding 0.4 µg/ml are associated with cholinergic illness. No changes in erythrocyte or plasma ChEs at average of 10 µg *p*-nitrophenol in urine were present. Hayes states that absorption of parathion is tolerated without illness and with little or no reduction of ChE activity, as long as the concentration of *p*-nitrophenol in urine does not rise much above 60 to 80 µg/h (2 ppm), assuming average urine excretion of 30 to 40 ml/h.[193]

Wolfe et al. demonstrated that *p*-nitrophenol excretion levels correlated with parathion exposure. It increased rapidly after spraying, then promptly decreased.[180] The average peak excretion level (about 170 µg/h) occurred 87 h after exposure began. Excretion became insignificant 5 to 8 d after exposure. Hayes et al. analyzed urine samples for the presence of alkyl phosphates in 22 pest control operators who applied vaponite, diazinon, and dursban; they used the alkyl phosphates as exposure indicators.[178] The air samples collected for 8-h exposures showed levels below the recommended threshold limits. Alkylphosphates were present in 96% of the 8-h specimens of the operators. The specific metabolite values detected in the samples were (presented as group mean $\pm$ 1 SD): 176.1 $\pm$ 9 µg/8h for DMP, 48.3 $\pm$ 58.6 for DEP, 5.6 $\pm$ 3.2 for DMTP, 8.3 $\pm$ 9 for DETP. Of the urine specimens collected, 96% contained DMP and/ or DEP. Low levels of DMTP and DETP were present in 28% of the samples. Urine samples collected from 2 persons not previously exposed to OP showed levels below the detection capability of the method. For employees of the company who were not actually involved in the application, very low levels (<15 µg/8 h) were found. The mean acetylcholinesterase value of the study group showed statistically significant plasma AChE inhibition (p <0.001). When compared to a control group RBC AChE values were not significantly different.

From the study of Ando et al. it appears that the cholinesterase activity is related to factors such as age, hemoglobin content, serum total cholesterol, transaminase (GTP) activity, and Broca index.[194]

The effects on serum cholinesterase of many diseases which affect hepatic function are well known. Neoplasma, parenchymal liver diseases, malnutrition, acute infections, and some anemias, all depress serum cholinesterase activity.[183]

About 3% of the population have genetically determined, reduced serum cholinesterase activity. This genetic variance is not associated with any increased susceptibility to acetylcholinesterase inhibition or poisoning by OP and carbamates.[183] This genetic variation can be identified with the determination of a dubicaine number.

M. DETERMINATION OF OP EXPOSURE INDICATORS
1. Cholinesterase Activity

Measuring cholinesterase activity is important in the preliminary examinations of persons exposed to occupational risk from pesticides. Cholinesterase activity must be tested before work begins, in order to evaluate the individual variations and determine the possible presence of a genetically determined deficiency of cholinesterase activity.[183,195,197]

Variation of ChE activity was studied by Cavanga et al.[27] Maximum variations for 4-month periods were 7.7% in erythrocyte ChE and 23.2% in plasma ChE.

As a hazard level, whole blood ChEA inhibited by 30% of the pre-exposure level was proposed.[195]

The report of the International Workshop on Epidemiological Toxicology of Pesticides Exposure proposed an acetylcholinesterase inhibition level more than 30% the pre-exposure level.[198] This requires remedial action including temporary removal from further exposure and appraisal of work situation; 20% ChE inhibition in group workers may be used as a test for exposure and an indication to sanitary failures.

TABLE 8

Selected Organophosphorous Insecticides Arranged According to Their Ability to Inhibit Either Plasma or Red Cell Cholinesterase in Man[1]

Plasma enzyme more inhibited	RBC enzyme more inhibited
Chlorpyrifos	Dimefox
Demeton	Mevinphos
Diazinon	Parathion
Dichlorvos	Parathion-methyl
Malathion	
Mipafox	
Trichlorphon	

Gage also proposed the same percent of ChEA inhibition as a safe threshold for both plasma and erythrocytes.[199]

The sensitivity of the method is of great importance. The following are the hazard levels of cholinesterase depression, suggested by the working group of WHO: (1) acetylcholinesterase activity reduced by 30% of the pre-exposure value requires repeated determination, after appropriate intervals, and appraisal of the general and individual work situation and (2) acetylcholinesterase activity reduced by 50% requires immediate action, including temporary removal from further exposure and appraisal of the work situation.[200]

WHO working group gives more details concerning recommendations for ChE monitoring with a tintometer, for operators exposed in malaria control operations:[201]

1. Weekly monitoring of blood ChEA
2. Enforcement of better adherence to precautionary measures when about 25% reduction of activity is found
3. Investigation to find out the most likely reason, leading to overexposure and allocation of a lighter spraying schedule for the following week at 37.5% reduction
4. Withdrawal from spraying and any other exposure to insecticide when 50% reduction in activity is found
5. Withdrawal period lasting until cholinesterase activity returns to normal, which may sometimes take more than 2 weeks
6. Better adherence to precautionary measures ensured by field supervisor when several persons have 12.5% ChEA reduction

The data from a survey of malaria control personnel in Pakistan and Haiti, presented by Miller and Shah, show that the tintometer method with a modified kit answers all the necessary criteria for determining cholinesterase activity in the field.[202]

Izmirova reports for the advantages of use of paper test in field conditions.[203] Other laboratory methods can be used if sample transportation is possible.

Both plasma/serum and RBC ChEA should be studied, because restoration of both is different and the inhibitory potential of various OPs for serum and RBC ChE is not equal (Table 8). When necessary, a debucaine number should be determined to rule out genetic deficiencies.

In the series of data sheets on pesticides (FAO, and WHO-VBC), the following indications concerning ChE levels are given:

	Normal levels	Action levels	Symptomatic levels
Plasma ChE	100%	50%	Variable
Erythrocyte ChE	100%	70%	40%

Hackathorn et al. validated a whole blood method for erythrocyte ChE.[204] They also proposed
baseline determination by 3 separate tests prior to exposure; where 70% of baseline values are
found, investigation and correction of exposure situation is needed. If 60% of baseline is
reached, removing workers from the exposure is recommended. This percentage is not based on
other investigations, but on calculation of the presumed ErChEA in the whole blood. So the
authors accept that 60% of the whole blood ChEA would correspond to 70% ErChEA. Normally
92% of the whole blood ChEA is due to Er if hematocrit is 50.

Vandekar[211] and Coye et al.[183] prefer determination of erythrocyte or blood ChE, which
correlates better to activity at nerve synapsis than plasma ChE.

Measurements of ChE assay procedures vary greatly, but the most satisfactory is that based
on the procedure of Elman et al.[205] A field procedure and kit for whole blood and plasma ChE
determination have been developed (WHO, 1984). Quick methods exist for determining ChE
in serum with paper tests — Izmirova, and for the colorimetric ChE determination of whole
blood — tintometer.[203]

In a comparative evaluation of the existing laboratory methods (electrometric, colorimetric,
titrimetric, tintometer and field spectrophotometric) Coye et al. determine the settings in which
each method is most useful.[183] The first two were suggested for surveillance and field research
in developed countries and for reference laboratories in developing countries. The last two were
proposed for field studies in developing countries and potential surveillance for field workers
in developed countries.

The Fifth International Workshop of the Scientific Committee on Pesticides of the International Association of Occupational Health on Field Exposure to Pesticides (1979) made the
following conclusion:

"There was no consensus on which types of cholinesterase should be measured. It appeared
that there is still not enough knowledge on the biological significance of plasma ChE, whole
blood ChE and erythrocyte ChE. The problem is particularly relevant in field work as one prefers
to test only one type of ChE under field conditions."[206] In many field conditions, procedures
using whole blood are more practical than those using separated erythrocytes. Quite commonly,
pseudo ChE is more sensitive to inhibitors.[207] Thus, if separation of plasma and erythrocytes is
possible prior to assay, an indication of exposure may be obtained by assay of pseudo ChE only.
Larsen and Hanel also found that the serum ChE is an adequate and sensitive method for
biologically monitoring OP exposure.[208]

Activity of A-esterase (phosphoryl-phosphatase), called paraoxonase by some authors, is
measured in order to determine worker sensitivity to OP.

2. OP Metabolites in Urine

Advances in gas chromatography and combined gas chromatography/mass spectrometry
(CG/MS) have made it possible to analyze the urine of exposed persons for the presence of
metabolites. It is usually necessary to preserve the sample by adding chloroform. This
concentrates or extracts the metabolite(s), and conducts them to suitably volatile derivatives that
can be detected by GC. In some cases, simpler and sensitive colorimetric tests are available for
screening the urine of exposed persons. Thus, 4-nitrophenol may be measured directly, in the
urine of workers exposed to parathion, as well as to methylparathion and chlorthion.[180]

3. Alkylphosphates in Urine (AP)

Urinary alkyl phosphate levels provide good information regarding the absorption and
excretion of OP pesticides. Analytical procedure is proposed by Shafic et al.[209] In a survey on
the general population of the United States for pesticide metabolites, Kutz et al. found substances
indicating OP exposure (Table 9).[210]

Coye et al. reviewed methods for determining alkylphosphates and phenolic substances.[211]
These authors quote studies demonstrating detection of urinary metabolites of organophosphates

TABLE 9
Pesticides and Metabolites Found in the General Population[210]

Chemical	Percent positive
Dimethyl phosphate (DMP)	76.7
Diethyl phosphate (DEP)	94.0
Dimethyl phosphothionate (DDETP)	70.8
Dimethyl phosphodithionate	38.2
Diethyl phosphodithionate (DEDTP)	0.4
Malathion	0.4

several days after exposure and in association with lesser exposures than those necessary to produce cholinesterase inhibition. A number of drawbacks and practical considerations have been identified in these methods however. Further studies are needed both to develop rapid nonsophisticated methods of analysis and demonstrate the value of AP as a monitoring test.

4. OP in Human Blood

Gas chromatography has been used successfully by Fournier and Sonnier to detect organophosphate insecticides in the blood samples of patients with acute poisoning and subjects with occupational exposure within 24 h.[212] The limit of detection in their study was 0.1 ng/ml. The authors related the difficulties in identification beyond 48 h to the high reactivity of these compounds.

5. EMG

Electroneuro-myography (ENM) appears to be an objective and sensitive test of neuromuscular function in OP-exposed persons.

This method was first used by Roberts and Wilson in the supervision of occupationally exposed organophosphate workers.[105]

Noninvasive surface electrodes are used for electromyographic studies.[106] The method requires special electrophysiological equipment and trained specialists. Some environmental factors affecting skin temperature may influence the results. Jager et al., Roberts and Trollope, and the Protocol for Epidemiological Studies on Neurotoxicity of Organophosphates (EURO/WHO) described the method in detail.[105,213]

Jager extensively reviewed 218 papers to throw light on the problem of neuropathy connected with OP exposure and the use of ENMG as a screening method for early detection of health impairments. He also reported results of his own and summarized the changes found as follows:[214]

1. A decrease in the voltage of muscle action potential, most values between 11 and 12 mV, in nonexposed controls, and wider ranges with values occasionally as low as 4 mV
2. A change in the pattern of the recorded electromyogram, indicating that not all muscle fibers react at the same time to the supramaximal stimulation (repetitive EMG activity in response to single stimuli)
3. A decrease in the conduction velocity of the nerve fibers

Jager's data on organophosphate workers indicate that the changes are reversible, but the literature he reviewed shows that in more advanced stages, which are accompanied by clinical signs and symptoms, this may not be always so.[214] Reversibility possibly depends on both the magnitude and the duration of exposure.

The slow recovery and the changes found are in accordance with a diagnosis of subclinical neuropathy. Neuropathy, no matter how slight, must be regarded as a more direct parameter of human health than a quickly reversible cholinesterase depression.

The author considers the parameter commonly used in the supervision of the workers exposed to organophosphate insecticides — the blood cholinesterase determination as a measure of exposure. ENMG changes, however, may be unrelated to depressions in ChE activity. Certainly there is a phase difference between the two; it may take days or weeks for ENMG changes to appear following overexposure, and it may take months to recover from it.

Cholinesterase determination and electromyography both have their own place in the medical supervision of organophosphate workers. Which one gives the first indication of overexposure probably depends on the timing of the test, in relation to the exposure, and the degree and duration of the exposure.

At present, ENMG-monitoring has a practical value in that it can supervise industrial hygiene and check the efficiency of protective and safety measures. It can also serve as another source of information on which the industrial medical director can base his clinical judgement. Quantick and Perry, as well as Violante and Roberts, suggest that periodical EMG checks for workers exposed to OP are more informative with regard to health effects.[215-216]

6. NTE

NTE measurement in peripheral lymphocytes was suggested as a biochemical test for delayed neuropathy caused by OP.[106,217] Bertoncin et al. have compared the NTE of lymphocytes and brain NTE. They concluded that these two enzymes are the same.[218] Recently, more researchers use these methods to monitor lymphocyte NTE in workers exposed to a potentially neurotoxic OP. As in the case of ChEA, the baseline (pre-exposure levels) of NTE will increase the biological sensitivity of the measurement. Variation of NTE values are 21.5% and are not age or sex dependent.

Lotti et al. interpreted the 40 to 60% levels of lymphocytic NTE inhibition as a warning that the exposure had approached the toxic threshold, and the 70 to 80% decrease from the pre-exposure value as a level that could produce neurotoxic response.[219] In animal studies designed to evaluate a new OP compound, the benefits of incorporating assays of NTE are obvious.[220] The possibility of monitoring exposed individuals by means of a human lymphocyte or platelet NTE activity is being explored by Lotti et al.[219]

N. OTHER TESTS

Hyperglycemia, found in acute intoxications, correlates well with the symptoms. Leucocytosis (with neutrophylia and eosinopenia) is found in acute intoxication, and leukopenia and anemia with thrombopenia are reported in chronic intoxication. Renal function should also be examined, including phosphorous reabsorption. Tests for changes in proteins (increased α-globulins) and serum enzyme activity (aspartate-aminotransferase, aldolase, alkaline phosphatase, SGOT, SGPT) may demonstrate some liver or other organ disturbances.

Salas and Christeva-Mircheva found leucocytosis and increased SRE parallel to decreased ChE activity in plasma.[220]

ECG investigations may also be useful. Functional impairments of the cardiovascular system have been found, such as: bradycardia, sinus arrhythmia, hypertonia, and ECG changes, including raised or inverted T wave and disappearing P wave.

Maroni gave a comprehensive review of the biological indicators of exposure in human beings.[221]

O. TREATMENT OF ORGANOPHOSPHATE INTOXICATION

First aid should be given as soon as possible, even by nonmedical but specially trained persons. It may also include atropine administration and other procedures, according to the state of the patient and the route of absorption. The purpose should be to minimize the absorption (remove contaminated chain, wash the skin with alkaline soap, and induce vomiting or gastric lavage with bicarbonate solution or activated charcoal), to support the vital functions (heart, respiration), and to start the specific pharmacological treatment.[1] Atropine is given i.v. beginning with 2 mg, at 15 to 30 min intervals. The dose and frequency varies in relation with the state of the patient, but it should maintain full atropinization (dilated pupils, flushed skin, dry mouth, etc.). Pulse rate should not exceed 120 per min. Cyanosis should be corrected first. In severe intoxications continuous infusion of atropine may be necessary.

Reactivators of ChE should be applied as soon as possible to avoid phosphorylated ChE aging. Pralidoxime is the most widely used oxime (1 g i.m. or i.v. 2 to 3 times daily, usually for 1 to 3 d). Diazepam is used to relieve CNS symptoms. Doses of 10 mg s.c. or i.v. are recommended. However, specific guidelines for the required amounts of atropine and pralidoxime and the duration of antidotal or other therapy are not available.[222] This is a question to be decided by the physician in each case.

To treat the severe poisoning of a person who ingested a terpentine solution of dicrotophos, Warringer et al. administered a total of 3911.5 mg of atropine i.v. over 16 d and 92 g of pralidoxime chloride i.v. over 23 d with successful therapeutic outcomes.[222] The daily dose of atropine reached 600 mg given at 15 min intervals, titrated to clinical response by continuous i.v. infusion. The every hour continuous infusion of pralidoxime was 0.5 g.

Nagler et al. reported a case of human poisoned with 10 g dimethoate responding favorably to prolonged combined hemoperfusion — hemodialysis initiated some 30 min after the ingestion.[223]

A patient with apnea, hypothermia, and shock after ingesting at least 150 mg of parathion as a 25% solution in methyl alcohol, has been successfully treated with obidoxime (1075 mg the first 3 d) and atropine (44.5 mg the first day and an average of 30 mg daily over the next 2 weeks).[81,82] Charcoal perfusion and exchange transfusion were carried out, as well as artificial respiration.

The high importance of skin cleaning should be noted. Magnesium oxide or other absorbents may be used.[221] Natrium bibarbonicum solution is one of the most usually applied means for skin cleaning because of its good availability and effectiveness.

VIII. PREVENTION

There is no internationally accepted permissible level for OPs in working environment. The WHO study group evaluated the information available on malathion and concluded that, in view of the recent data on the role of impurities in malathion toxicity, there is no specific basis for such a recommendation. A reduction of 30% of ChE activity should be accepted as a biological limit.[225]

Existing permissible levels for OP in the USA and USSR are shown in Table 10.[226-227]

Many joint FAO/WHO meetings have evaluated existing information on OP toxicity, and acceptable daily intakes have been recommended (Table 11).[228-230]

Dermal absorption is very important during occupational exposure.[231] Preventive measures should be taken to avoid skin contact and ensure immediate cleaning of contaminated skin.[17,221,231]

IX. CONCLUSION

Organophosphorous compounds are one of the most frequently used group of pesticides in the world. In agriculture, forestry, and public health they find application as insecticides, fungicides, herbicides, and raticides. More than 100 individual compounds from this group are well known as pesticides.

They enter the body easily via skin during occupational exposure. Organophosphorous pesticides are very dangerous due to their high acute toxicity (LD_{50} from mg to g/kg/b.w.); however, they have a low potential for cumulation.

TABLE 10
Permissible Levels for Some Pesticides (in mg/m³)

Pesticides	USA			U.S.S.R
	TWA	STEL	IDLH	
Abate				0.5
Afugan				0.05
Amidophos				0.5
Amiphos				0.5
Anthio				0.5
Azynphos-methyl	0.2	0.6		
Basudin				0.2
Baytex				0.3
Bromophos				0.5
Carbophos				0.5
Chlorophos				0.5
Cyanox				0.3
DDVP				0.2
Demeton	0.1	0.3	20	
Diazinon	0.1			
Dibrom				
Dichlorvos	1	3	200	0.5
Dicrotophos	0.25			
Fenamiphos	0.1			
Fensulfothion	0.1			
Fenthion	0.2			
Iodofenphos				
Kilval			0.5	
Malathion	10–15		5000	
Methafos				0.1
Methylacetophos				0.1
Methylmercaptophos				0.1
Methylethylthiopfos				0.1
M,81 (o-o-dimethyl-S-				0.03
ethylmercaptoethyl-				0.1
dithiophosphate)				
Methylparathion	0.2	0.6		
Mevinphos	0.1	0.3		
Monocrotophos	0.25			
Naled	3	6		
Octamethyl				0.02
Parathion	0.1/0.05	0.3	20	0.05
Phosalon			0.05	
Phosphamide				0.5
Phtalophos				0.3
Sayphos				1
Trichlorometphos				0.3
Trolen				0.3
Valexon				0.1

Three main biochemical reactions are responsible for OP effects: inhibition of cholinesterase activity (ChE inhibition), inhibition of neuropathy target esterase (NTE) and release of alkyl groups attached to the phosphorus atom, and alkylation of macromolecules RNA and DNA.

ChE inhibition is their specific effect and principal mechanism of action. Accumulation of acetylcholine at the nerve ending produces muscarinic and nicotinic effects. Central nervous system is effected by direct impact on the choline receptors of the nerve cells, as well.

Acute intoxications with OP are very serious health problems. The percentage of lethal cases is very high in accidental and intentional intoxication.

The clinical picture is dominated by cholinergic symptoms of a different grade. The most common of them are headache, salivation, myosis, lacrimation, dyspnea, nausea, vomiting,

TABLE 11
Acceptable Daily Intake (ADI)

Pesticide	ADI in mg/kg/b.w.
Bromophos	0.006
Bromophos ethyl	0.003
Chlorpyrifos	0.0015
Chlorpyriphos methyl	0.01
Coumaphos	0.0005
Demeton	0.005
Diazinon	0.002
Ethion	0.0005
Etrimphos	0.003
Fenthion	0.001
Fenamiphos	0.0003
Fenchlorphos	0.01
Fenitrothion	0.003
Fensulfothion	0.0003
Formothion	0.02
Isofenphos	0.005
Malathion	0.02
Metacrifos	0.0003
Methidothion	0.05
Mevinphos	0.0015
Metamidophos	0.006
Monocrotophos	0.0006
Omethoate	0.0003
Parathion methyl	0.001
Phentoate	0.003
Phosmet	0.02
Phosalone	0.006
Phosphamidon	0.0005
Thiophanate-methyl	0.08
Triazephos	0.0002
Trichlorfon	0.01
Vamidothion	0.0003

sweating, fatigue, giddiness, tension, anxiety, muscular fascillation, tremor, ataxia, convulsion, lung edema, and coma. Together with the symptomatic treatment, an adequate specific therapy with cholinolytics (atropine) and ChE reactivators is applied. The intoxicated person may recover completely.

Chronic intoxications are rather rare. In epidemiological studies, most frequently nervous system, liver, renal, skin, cardiovascular, hemopoietic, and respiratory disturbances are reported.

Delayed neuropathy due to NTE inhibition, subsequent axon degeneration, and demyelinization may develop after contact with some OPs. Distal paralysis usually occurs three weeks after the beginning of the intoxication. Electroneuromyography is used for diagnosis. Behavioral deviations and EEG changes are often related to central nervous system impairment due to OPs.

ChE activity is used as a specific test for exposure and is included as a principal means for biological monitoring of agricultural workers. Recently, NTE measurement in peripheral lymphocytes has been suggested as a biochemical test for delayed neuropathy by OP. Alkylphosphates in urine provide good information regarding absorption and excretion of OP pesticides.

Reentry intoxications are a new problem and require specific legislation.

REFERENCES

1. Organophosphorus Insecticides. A general Introduction, Env. Health Criteria No. 63, World Health Organization, Geneva, 1986.
2. **Wolfe, H., Durham, W. F., Armstrong, J. F.,** Exposure of workers to pesticides, *Arch. Environ. Health,* 14, 622, 1967.
3. **Maibach, I. H., Feldmann, R. J., Milby, T., and Serat, W. F.,** Regional variation in percutaneous penetration in man, *Arch. Environ. Health,* 23, 208, 1971.
4. **Davies, J. F., Barquest, A., Free, V., Hagie, R., Morgate, C., Sonnenborn, R., and Vaclavek, C.,** Human pesticides poisonings by a fat-soluble organophosphate insecticide, *Arch. Environ. Health,* 30, 608, 1975.
5. **Georgiev, G.,** Toxicokinetics of pesticides, in *Atlas of Toxicokinetics,* Popov, T., Zapryanov, Z., Bentchev, I., and Georgiev, G., Eds., Med. Fizkult., Sofia, 150, (in Bulgarian).
6. **Aldridge, W. H. and Magos, L.,** *Carbamate, Thiocarbamate, Dithiocarbamates,* CEC, Luxemburg, 1978.
7. WHO, TRS No 560, Chemical and biochemical methodology for the assessment of hazards of pesticides for man, World Health Organization, Geneva, 1975.
8. **Aldridge, W. H. and Reiner, E.,** *Enzyme Inhibitors as Substrates,* Nord Holland, Amsterdam No. 328.
9. WHO TRS No 356, Safe use of pesticides in public health: 16th report of the WHO Expert Committee on Insecticides, 1967.
10. WHO TRS No. 634, Safe use of pesticides: 3th report of the WHO Expert Committee on Vector Biology and Control, 1979.
11. **Johnson, M. K.,** Delayed neurotoxic action of some organophosphorus, *Br. Med. Bull.,* 25, 3, 231, 1969.
12. **Lotti, M. and Johnson, M. K.,** Organophosphate neurotoxicity. Neurotoxic esterase in human nervous tissue, Abstract, *Toxicol. Appl. Pharmacol.,* 28, 5, 1979.
13. **Abdel Aal, H. M. A., El Hawari, M. P. S., Kamel, H., Aboel-Khabk, M. K., and El-Diwany, M. K.,** Blood cholinestereases, hepatic, renal and hemopoietic functions in children receiving repeated doses of dipterex, *J. Egypt. Med. Assoc.,* 53, 265, 1970.
14. **Beneyt, P., Lebrun, A., Cerf, J., Dierickx, J., and Degroote, R.,** Etude de la toxicité pour l'homme d'un insecticide organophosphoré, *Bull. Org. Mond. Santé,* 24, 465, 1961.
15. **Derache, R.,** Rep. Organophosphorus Pesticides - Criteria (dose) effect relations, for Organophosphorous Pesticides, Rep. of the Working Group of Experts for the CEC, Pergamon Press, 1977.
16. **Laws, I., Morales, F., Hayes, W., and Joseph, C.,** Toxicology of abate in volunteers, *Arch. Environ. Health,* 14, 2, 289, 1967.
17. **Kagan, J.,** *Toxicology of Organophosphorous Pesticides,* Medizina, Moscow, 1977, (in Russian).
18. Shradan (Octamethyl) 82, IRPTC Sc. Rev. of Soviet Literature, CIP, GKNT, Moscow, 1985.
19. Malathion 8, IRPTC Sc. Rev. Soviet Literature, CIP, GKNT, Moscow, 1982.
20. Parathion 10, IRPTC Sc. Rev. Soviet Literature, CIP, GKNT, Moscow, 1982.
21. DDVP 79, IRPTC Sc. Rev. Soviet Literature, CIP, GKNT, Moscow, 1984.
22. **Kawal, M., Yoshida, M., Iyatumi, A., Koyeama, M., Kaneko, Y.,** *Jpn. J. Publ. Health,* 29, 4, 171, 1982.
23. **Soliman, S. A., El-Sebae, A. H., and El-Fiki, S.,** Occupational effect of pholan insecticides on spraymen during field exposure, *J. Environ. Sci. Health,* 14, 1, 27, 1979.
24. **Lu, Y. P., Lu, P. K., Xue, S. Z., and Gu, X. G.,** Investigation on the chronic effects of dipterex in occupational exposure, *Med. Lav.,* 75, 5, 376, 1984.
25. **Wolfe, H., Armstrong, F., Staiff, C., and Comer, M.,** Exposure of spraymen to pesticides, *Arch. Environ. Health,* 25, 6, 29, 1972.
26. **Wolfe, H, Staiff, D. C., and Armstrong, J. E.,** Exposure of pesticides formulating plant workers to parathion, *Bull. Environ. Contam. Toxicol.,* 21, 340, 1978.
27. **Cavanga, G., Locati, G., and Vigliani, E. C.,** Clinical effect of exposure to DDVP (Vapona) insecticide in hospital wards, *Arch. Environ. Health,* 19, 112, 1969.
28. **Cavanga, G. and Vigliani, E. C.,** Problemes d'hygiene et de securité dans l'emploi de vapona comme insecticide dans le locaux domestiques, *Med. Lav.,* 11, 8, 409, 1970.
29. **Shin, J. H., Wu, Z. Q., Wang, Y., Zang, Y. L., Xue, S. Z., and Gu, X. Q.,** Prevention of acute parathion and demethion poisoning in farmers around Shanhai, *Scand. J. Work Environ. Health,* 11, 49, 1985.
30. **Backer, E., Zack, M., Miles, J., Alferman, I., McWilson, W., Robbin, R., Miller, S., and Teeters, W.,** Epidemiologic malathion poisoning in Pakistan malaria workers, *Lancet,* 7, 31, 1978.
31. **Feldman, R. J. and Maibach, H. T.,** Percutaneous penetration of some pesticides and herbicides in man, *Toxicol. Appl. Pharmacol.,* 28, 126, 1974.
32. **Davis, J. E., Stevens, E. R., Staiff, D. C., and Butler, L. C.,** Potential exposure of apple thinners to phosalone, *Bull. Environ. Contam. Toxicol.,* 29, 5, 592, 1982.
33. **Devine, J. M., Kinoshita, G. B., Peterson, R. P., and Picard, G. L.,** Farm workers exposure to terbufos (phosphoro-dithioic acid, *S*-(tertbutylthio)methyl O,O-diethyl ester) during planting operations of corn, *Arch. Environ. Contam. Toxicol.,* 15, 1, 113, 1986.

34. **Zakurdaev, V. V.,** Toxic myopathy at acute intoxications with OP, *Voen. M. Zh.,* 10, 49, 1986, (in Russian).
35. **Golden, A., Rubinstei, A., Bradlov, B., Path, D., Path, M., and Elliott, G.,** Malathion poisoning with special reference to the effect of cholinesterase inhibition on erythrocyte survival,*N. Engl. J. Med.,* 271, 25, 1289, 1964.
36. **Favre, S.,** Thése pour le doctorat en medicine, Paris, 1966, 752.
37. **Gervais, P., Frejaville, J., and Efthymiou, M.,** Intoxication par un organophosphoré, IMMEX, 4, 385, 1967.
38. **Gupta, O. and Patel, D.,** Diasinon poisoning, a study of 60 cases, *J. Assoc. Phys. Ind.,* 16, 457, 1968.
39. **Tevfik, O.,** A propos de plusieurs cas d'intoxications par un pesticide à base de parathion, le corthion, *La Press Médicale,* 67, 17, 1959.
40. **Bledsoc, F. and Seymour, E.,** Acute poisoning oedema associated with parathion poisoning, *Radiol.,* 103, 53, 1972.
41. **Cattle, D.,** A case of organophosphorous poisoning, *Centr. Afr. J. Med.,* 19, 3, 52, 1973.
42. **Gaultier, M., Fournier, E., Gervais, P., and Gorceix, A.,** Traitement par le N'-Methyl-alpha-pyridil-aldoxine d'une intoxication grave par un insecticide organophosphoré (parathion), *Arch. Mal. Prof.,* 22, 1, 55, 1961.
43. **Wickoff, D., Davies, J., Barquest, A., and Davis, J.,** Diagnostic and therapeutic problems of parathion poisoning, *Ann. Intern. Med.,* 68, 875, 1968.
44. **Schachter, A. and Garda, M.,** Contributa la stuiul intoxication cu compusi organofosforici, *Med. Int.* (Bucuresti), 18, 3, 285, 1966 (in Rumanian).
45. **Taberschaw, J. and Cooper, W.,** Sequelae of acute organic phosphate poisoning,*J. Occup. Med.,* 8, 1, 5, 1966.
46. **West, J.,** Sequelae of poisoning from phosphate ester pesticides, *Int. Med. Surg.,* 37, 7, 538, 1968.
47. **Hartwell, W. and Hayes, G.,** Respiratory exposure to organic phosphorous insecticides,*Arch. Environ. Health,* 11, 4, 564, 1965.
48. **Faerman, I.,** Some questions of the clinic of intoxications by toxic organophosphorous chemicals,*Gig. Primen. Toksikol. Pestits. Klin. Otravlenii,* 542, 1985, (in Russian).
49. **Kudo, H.,** Practical clinical report of intoxication caused by organophosphorous in rural villages around Hirosaki in Aomori Prefecture, *I ACRAM,* 188, 1973.
50. **Nakazawa, T.,** Chronic organophosphorous intoxication in females, *Proc. I ACRAM,* 190, 1974.
51. **Bogusz, M.,** Studies on the activity of certain enzymatic systems in agriculture workers exposed to the action of organophosphorous insecticides, *Wiad. Lek.,* 23, 786, 1968.
52. **Grech, J. L.,** Alternations in serum enzymes after repeated exposure to malathion,*Br. J. Ind. Med.,* 22, 67, 1965.
53. **Burseva, L.,** On the chronic effect of plant protection chemicals on the functional state of liver in people working with them, *Higiena i zdraveopazvane,* 9, 481, 1966, (in Bulgarian).
54. **Gaaz, J., Poser, W., and Erdmann, W.,** Research on liver toxicity of nitrostigmin (parathion E 605) in the perfused rat liver, *Arch. Toxicol.,* 33, 31, 1974.
55. **Karimov, M.,** Free amino acid content in serum of persons occupationally exposed to pesticides, *Gig. Primen. Toksikol. Pestits. Klin. Otravlenii,* 548, 1968, (in Russian).
56. **Kay, R.,** Occupational cancer risks for pesticide workers, *En. Res.,* 7, 243, 1974.
57. **Davies, E., Mann, B., and Tocci, M.,** Renal tubular dysfunction and amino acid disturbances under conditions of pesticide exposure, *Ann. N.Y. Acad. Sci.,* 160, 322, 1969.
58. **Orlando, E. and Raffi, G.,** Galluppi Fenomini di assueffazione as composti organofosforici,*Min. Med.,* 59, 78, 4090, 1968.
59. **Tocci, M., Mann, B., Davies, I., Edmundson, W. F.,** Biochemical differences found in persons chronically exposed to high levels of pesticides, *Ind. Med.,* 36, 6, 1969.
60. **Faerman, I.,** Some problems of the clinic of intoxications by organophosphorous pesticides, *Gig. Primen. Toksikol. Pestits. Klin. Otravlenii,* 542, 1965, (in Russian).
61. **Metcalf, R. and Holmes, J. H.,** EEG, psychological and neurological alternations in humans with organo-phosphorous exposure, *Ann. N.Y. Acad. Sci.,* 160, 357, 1969.
62. **Christophere, R. J.,** Hematological effect of pesticides, *Ann. N.Y. Acad. Sci.,* 160, 155, 1969.
63. **Kaloyanova-Simeonova, F.,** *Pesticides Toxic, Action and Prophylaxis,* Bulgarian Academy Science, 1977.
64. **Matsushita, T., Aoyama, K., Yoshima, R., Rujita, Y., and Ueda, A.,** Allergic contact dermatitis from OP insecticides, *Ind. Health,* 23, 145, 1985.
65. **Milby, P. and Epstein, W.,** Allergic contact sensitivity to malathion, *Arch. Environ. Health,* 9, 434, 1964.
66. **Ganelin, R., Gueto, C., and Mail, G.,** Exposure to parathion effect on general population and asthmatics, *JAMA,* 188, 807, 1964.
67. **Davignon, L., Pierre, I., Charest, G., and Torrangeau, F.,** A study of the chronic effects of insecticides in man, *Can. Med. Assoc. J.,* 92, 3, 597, 1965.
68. **Gardner, A. and Iverson, R.,** The effect of aerially applied malathion on an urban population, *Arch. Environ. Health,* 16, 823, 1968.
69. **Edmundson, W. and Davies, J.,** Occupational dermatitis from naled. A clinical report, *Arch. Environ. Health,* 15, 89, 1967.
70. **Ercecovic, C. D.,** Relationship of pesticides to immune responses, *Fed. Proc.,* 32, 2010, 1973.
71. **Johnson, M. K.,** Delayed neurotoxicity tests of organophosphorous ester: a proposed protocol integrating

neuropathy target esterase (NTE) assays with behaviour and histopathology tests to obtain more information more quickly from fewer animals, in *Proc. of Int. Conf. on Environ. Hazards of Agrochemicals in Developing Countries,* El Sebae, A. H., Ed., Alexandria University, 1, 747, 1987.

72. **Aldridge, W. H. and Johnson, M. K.,** Side effects of organophosphorous compounds; delayed neurotoxicity, *Bull. WHO,* 44, 259, 1971.

73. **Vilanova, E., Johnson, M. K., and Vicedo, J. L.,** Interaction of some unsubstituted phosphoramidate analogs of methamidophos (O,S-dimethyl phosphorothioamidate) with acetylcholinesterase and neuropathy target esterase of hen brain, *Pestic. Biochem. Physiol.,* 28, 1987.

74. **Johnson, M. K.,** Receptor as enzyme: the puzzle of NTE and organophosphate-induced delayed polyneuropathy, in *Trends in Pharmacological Sciences,* (in press).

75. **Senanayake, N. and Johnson, M. K.,** Acute polyneuropathy after poisoning by a new organophosphate insecticide, *N. Engl. J. Med.,* 306, 3, 155, 1982.

76. **Lotti, M., Becker, C. E., Aminoff, M. J., Woodro, J. E., Seiber, J. M., Talcott, R. E., and Richardson, R. J.,** Occupational exposure to the cotton defoliants DEF and merphos, *J. Occup. Med.,* 25, 517, 1983.

77. **Bidstrup, B., Bonnel, A. J., and Bleckett, B.,** Paralysis following poisoning by a new organic phosphorus insecticide (mipafox), *Br. Med. J.,* 1, 1068, 1953.

78. **Murphy, S.,** Pestcides, in *Toxicology: a Basic Science of Poisons,* Doull, J., et al., Eds., McMillan, New York, 1980, 357.

79. **Osterloh, J., Lotti, M., and Pond, C. M.,** Toxicological studies in a fatal overdose of 2-4-D, MOPP and chlorpyrifos, *J. Anal. Toxicol.,* 7, 125, 1983.

80. **Hodgson, M. J., Block, G. D., and Parkinson, D. R.,** Organophosphate poisoning in office workers, *J. Occup. Med.,* 28, 6, 435, 1986.

81. **De Jager, A. E. J., Van Weerdon, T. W., Houtoff, H. J., and De Monchy, J. G. R.,** Polyneuropathy after massive exposure to parathion, *Neurology,* 31, 603, 1981.

82. **De Jager, A. E. J., Van Weerdon, T. W., and Monchy, J. G. R.,** Reply to letter, *Neurology,* 32, 218, 1982.

83. **Xintaras, C., Burg, J. R., Tanake, S., Lee, S. T., Johnson, B. L., Cottrill, C. A., and Bender, J.,** Occupational exposure to leptophos and other chemicals, U.S. Department of Health, Education and Welfare, Cincinnati, Ohio, 1978.

84. **Anon.,** Evaluation of some pesticide residues in food, FAO Plant Production and Protection Paper 13, Suppl., Food and Organization/World Health Organization, Rome, 1979.

85. **Shiraishi, S., Goto, I., Yamashita, Y., Onishi, A., and Nagao, H.,** Dipterex polyneuropathy, *Neurol. Med.* (Japan), 6, 34, 1977, (English summary).

86. **Hirons, R. and Johnson, M. R.,** Clinical and toxicological investigations of a case of delayed neuropathy in man after acute poisoning by an organophosphorous pesticide, *Arch. Toxicol.,* 40, 279, 1978.

87. **Johnson, M. K.,** Do trichlorphon and/or dichlorvos cause delayed neuropathy in man or in test animals?, *Acta Pharmacol. Toxicol. Scand.,* 49, 87, 1981.

88. **Jerdzejowska, H., Rowinska-Marinska, K., and Hoppe, B.,** Neuropathy due to phytosol (agritox): report of case, *Acta Neuropathol.,* 49, 163, 1980.

89. **Williams, J. L.,** Poisoning by organophosphate insecticides, analysis of 53 human cases with regard to management and long treatment, *Acta Med. Milit. Belg.,* 134, 7, 1981.

90. **Xintaras, C. and Burg, J. R.,** Screening and prevention of human neurotoxic outbreaks, issues and problems, in *Clinical and Experimental Neurotoxicology,* Spenser, P. S. and Schaumburg, N. N., Eds., Williams and Wilkins, Baltimore, 1980, 663.

91. **Lotti, M. and Marretto, A.,** Inhibition of lymphocyte neuropathy target esterase predicts the development of organophosphate neuropathy in man, *Human Toxicol.,* 5, 114, 1986.

92. **Petry, C.,** Organic phospate insecticide poisoning, *Amer. Med.,* 22, 467, 1958.

93. **Healy, J. K.,** Ascending paralysis following malathion intoxication: a case report, *Med. J. Austr.,* 46, 765, 1959.

94. **Petry, H.,** Polyneuritis from E 605, *Arbeitsmed. Arbeitssch.,* 1, 239, 1973.

95. **Fisher, J. R.,** Guillain-Barre syndrome following organophosphate poisoning, *J. Amer. Med. Assoc.,* 238, 1950, 1977.

96. **Curtis, J. P., Develay, P., and Hubert, J. P.,** Late peripheral neuropathy due to an acute voluntary intoxication by organophosphoric compounds, in *International Congress on Neurotoxicology, Varese, Italy, 27-30 Sept.,* Pergamon Press, Oxford, 1979.

97. **Lotti, M., Ferrara, S., Garoldi, S., and Sinigalia, F.,** Enzyme studies with human and hen autopsy tissue suggest omethoate does not cause delayed neuropathy in man, *Arch. Toxicol.,* 48, 248, 1981.

98. **Rayner, M., Popper, J., Caravalho, E., and Hurov, R.,** Hyporeflexia in workers chronically exposed to organophosphate insecticides, *Res. Commun. Chem. Pathol. Pharmacol.,* 4, 595, 1972.

99. **Anon.,** Organophosphorous pesticides. An Epidemiological study, World Health Organization Regional Office for Europe, Copenhagen, 1987.

100. **Kaloyanova, F. P.,** Health effects of low level exposure, nerotoxicity of organophosphorus compounds: background literature review, World Health Organization Regional Office for Europe, Copenhagen, 1982 (unpublished document).

101. **Batora, I.,** Pozdna neurotoxicii sa organickych zlucenin fosfor, *Prac. Lek.,* 32, 8, 286, 1980.

102. **Seppaleinen, A. M. H.,** Neurophysiological approaches to the detection of early neurotoxicity in humans, *CRC Critical Review in Toxicology,* CRC Press, Boca Raton, FL, 1988, 245.

103. **Johnson, B. L.,** Ed., *Prevention of Neurotoxic Illness in Working Population,* John Wiley & Sons, New York, 1987.

104. **Roberts, D. V. and Wilson, A.,** Monitoring biological effects of anticholinesterase pesticides, 8th Int. Conf. on Toxicology and Occupational Medicine, Miami, 1973, 479.

105. **Jager, K. W., Roberts, D. V., and Wilson, A.,** Neuromuscular function in pesticide workers, *Br. J. Ind. Med.,* 27, 273, 1970.

106. **Dudek, B. R., Barth, M., Gephart, J., Huggins, J., and Richardson, R. J.,** Correlation of brain and lymphocyte neurotoxic esterase inhibition in adult hen following dosing with neurotoxic compounds, *Toxicol. Appl. Pharmacol.,* 48, A198, 1979.

107. **Roberts, D. V.,** EMG voltage and motor nerve conduction velocity in organophosphorous pesticide factory workers, *Int. Arch. Occup. Environ. Health,* 36, 267, 1976.

108. **Drenth, H., Ensberg, I., Roberts, D., and Wilson, A.,** Neuromuscular function in agricultural workers using pesticides, *Arch. Environ. Health,* 25, 396, 1972.

109. **Roberts, D. V.,** A longitudinal EMG study of 6 men occupationally exposed to organophosphorous compounds, *Int. Arch. Occup. Environ. Health,* 38, 221, 1977.

110. **Verberk, M. M. and Salle, H. J. A.,** Effects on nervous function in volunteers ingesting mevinphos for one month, *Toxicol. Appl. Pharmacol.,* 42, 351, 1977.

111. **Jusić, A., Jurenić, D., and Milić, S.,** EMG neuromuscular synapse testing and neurological findings in workers exposed to organophosphorous pesticides, *Arch. Environ. Health,* 35, 3, 168, 1980.

112. **Hussain, M. A., Oloffs, P. C., Blatherwick, F. J., Gaunce, A. P., and Mackenzie, C. J. G.,** Detection of incipient effects of anticholinesterase insecticides in rats and humans by EMG and cholinesterase assay, *J. Environ. Sci. Health,* 316, 1, 1981.

113. **Karczmar, A. G.,** Acute and long-lasting central actions of organophosphorous agents, *Fundam. and Appl. Toxicol.,* 4, S1, 1984.

114. **Gershon, S. and Shaw, F.,** Psychiatric sequelae of chronic exposure to organophosphorous insecticides, *Lancet,* 1, 371, 1961.

115. **Dille, J. and Smith, P.,** Central nervous system effects of chronic exposure to organophosphorous insecticides, *Aerosp. Med.,* 9, 765, 1964.

116. **Monov, A.,** Some peculiarities in the clinic and treatment of parathion coma, *Savremena Med.,* 3, 16, 148, 1965, (in Bulgarian).

117. **Kovarik, J. and Sercl, M.,** The influence of the organophosphate insecticides on the nervous system of man, in 15th Int. Congr. on Occupational Health (Vienna), Abstracts, 6, 205, 1966.

118. **West, J.,** Sequelae of poisoning from phosphate ester pesticides, *Int. Med. Surg.,* 37, 7, 538, 1968.

119. **Maizlish, N., Schenker, M., Weisskopf, C., Seiber, J., and Samuels, S.,** A behavioural evaluation of pest control workers with short-term, low-level exposure to the organophosphate diazinon, *Amer. J. Ind. Med.,* 12, 153, 1987.

120. **Levin, H. S. and Rodnitzky, L.,** Behavioural effect of organophosphate pesticides in man, *Clin. Toxicol.,* 9, 3, 391, 1976.

121. **Maizlish, N.,** Neurobehavioural effect of organophosphate pesticides, excerpted from, Eskenaszi, B. and Maizlish, N., Effects of occupational exposure to chemicals on neurobehavioural functioning, in *Neuropsychological Disorders in Medical Illness,* Tartar, R. E., Van Thiel, D. H., and Edwards, K. L., Eds., Plenum Press, New York, 1987.

122. **Reich, G. and Berner, W.,** Aerial application accidents, *Arch. Environ. Health,* 17, 5, 776, 1968.

123. **Wood, W., Gabica, J., Brown, H., Watson, M., and Bonson, W.,** Implication of organophosphate pesticide poisoning in the plane crash of a duster pilot, *Aerosp. Med.,* 42, 10, 1111, 1971.

124. **Richter, E. D., Gribetz, B., Krasna, M., and Gordon, M.,** Heat stress and aerial spray pilots, in *Studies in Environmental Science. 7 Field worker exposure during pesticide application. I.,* Tordoir, W. E. and Van Heemstra, E. A., Eds., 1980, 129.

125. **Durham, W., Wolfe, H., and Quinby, G.,** Organophosphorus insecticides and mental alertness, *Arch. Environ. Health,* 10, 55, 1965.

126. **Brower, M., Goodman, E., and Sim, V.,** Some behavioural changes in man following anticholinesterase administration, *J. Nerv. Ment. Diseases,* 138, 383, 1964.

127. **Kaloyanova, F., Verginva, T., Dincheva, E., et al.,** Epidemiological study on the health effect from organnosphorous compounds, in *Problemi na Higienata,* Centre for Scientific Information on Medicine and Health, Sofia, 1989, 35, (in Bulgarian).

128. **Mellerio, F.,** *L'electroencephalographie dans les Intoxications Aigues,* Masson and Cie, Paris, 1969, 204.

129. **Dufy, F., Burgchfiel, J., Bartels, P., Goan, M., and Sini, V.,** Long-term effects of an organophosphate upon human electroencephalogramme, *Toxicol. Appl. Pharmac.,* 47, 161, 1979.

130. **Plestina, R. and Piukovic-Plestina, M.,** Effects of anticholinesterase pesticides on the eye and vision, *Crit. Rev. Toxicol.,* Dec., 1978.

131. **Revzin, A. M.,** Effects of organophosphate pesticides and other drugs on subcortical mechanisms of visual integration, *Aviation, Space and Environ. Med.,* June, 627, 1976.

132. IARC monographs on the evaluation of the carcinogenic risk of chemicals to humans, Miscelleneous Pesticides, 30, World Health Organization/International Agency for Research on Cancer, 1983.

133. **Kaloyanova, F.,** Pesticides toxic action and prevention, Publ. House of the Bulg. Acad. Sci., Sofia, 1977.

134. **Thiodet, J., Massonnet, J., Milhaud, M., Colonna, P., and Tonati, E.,** Intoxication par insecticides á base de parathion. Intérés des méthodes de réanimation respiratoire dans le traitment des forms graves, Sté Médicale des Hôpitaux d'Alger A. M., Février, 1960, 1983.

135. **Carini, R., Massair, L., and Quersi, V.,** The role of pathological antecedents in poisoning from parathion, *Folia Med.,* 50, 504, 1957.

136. **Canivet, M.,** Thèse pour le doctorat en médicine, Hop. Fern. Widal, Paris, 1964, 135.

137. **Arterberry, J., Durham, W., and Elliott, I.,** Exposure to parathion. Measurement by blood cholinesterase level and urinary paranitrophenol excretion, *Arch. Environ. Health,* 3, 476, 1961.

138. **Roan, C., Morgan, D., Cook, N., and Pascal, E.,** Blood cholinesterases, serum parathion concentrations and urine *p*-nitrophenol concentrations in exposed individuals, *Bull. Environ. Contam. Toxicol.,* 4, 6, 1969.

139. **Milthers, E., Clemesen, C., and Nime, N.,** Empoisonnement par les composés organophosphorés, *Dan. Med. Bull.,* 10, 4, 122, 1963.

140. **Kalovanova-Simeonova, F. and Fournier, F.,** Les pesticides et l'homme, Coll. de Méd. légale et toxicol. médicale, Masson & Cie, 1971, 166.

141. **Rissel, P., Duc, M., Dupres, A., and Rover, R.,** Intoxications aiguës par un organophosphoré. Effets du contrathion sur les troubles cardiaques et les activités cholinésterasiques, globulaires et tissulaires, *Bull. Méd. Lég. Toxicol. Méd.,* 8, 5, 398, 1965.

142. **Eitzamn, D. and Wolfson, S.,** Acute parathion poison in children, *Am. J. Dis. Child.,* 114, 142, 1967.

143. **Wakasuki, T.,** The actual state of pesticide poisoning in farmers in Japan, in 3rd Congr. Int. Med. Rur., Bratislava, 1966, 78.

144. **Ishikawa, S., Kimigoshi, W., and Massova, S.,** Organophosphorous pesticide intoxication in chronic cases, *Brain Nerve,* 24, 387, 1972.

145. **Rosin, D.,** Hygiene of work in cultivation of cotton with application of some organophosphorous pesticides, *Medicina Tashkent,* 1970, 36.

146. **Hegazy, M.,** Poisoning by Meta-Systox in spraymen and in accidentally exposed patients, *Br. J. Ind. Med.,* 22, 230, 1965.

147. **Tilsner, V.,** Kasuistischer Beitrag über den protrachierten Verluaf einer Pflanzenschutzmittel, Vergiftung, *Arzneim.-Forsch.,* 20, 5, 272, 1966.

148. **Bedhead, J.,** Poisoning on the farm. Report of a case of organophosphorous poisoning, *Lancet,* 1, 686, 1968.

149. **Fournier, E.,** Preuves clinic et biolgiques des intoxications par produits de maison. Rapport de séance pléniére UPAC, Prague 1968, *Pure Appl. Chem.,* 18, 151, 1969.

150. **Demidenko, M. N. and Mirgiiazova, M. G.,** The combined biological effects of high air temperature and the pesticide anthio, *Gig. Sanit.,* 7, 18, 1974.

151. **Grech, J.,** Alteration in serum enzymes after repeated exposure to malathion, *Br. J. Ind. Med.,* 22, 1, 67, 1965.

152. **Sellassie, M. and Lester, F.,** Malathion poisoning, *Eth. Med. J.,* 9, 205, 1971.

153. **Ramu, A., Slonim, A., and Eyal, F.,** Hyperglycemia in acute malathion poisoning, *Is. J. Med. Sci.,* 9, 631, 1973.

154. **Stevens, J.,** Effect of malathion on hepatic microsomal metabolism of the male mouse, *Pharmacol.,* 11, 330, 1974.

155. **Markovitz, J. S., Gutterman, E. M., and Link, D. G.,** Self-reported physical and psychological effects following a malathion pesticide incident, *J. Occup. Med.,* 28, 5, 377, 1986.

156. **Vandekar, M.,** Observations on the toxicity of carbaryl, folithion and 3-isopropylphenyl *N*-methylcarbamate in village scale trial in Southern Nigeria, *Bull. WHO,* 33, 1965.

157. **Menz, M., Leutkemier, H., and Sachsse, K.,** Long-term exposure of factory workers to dichlorvos (DDVP) insecticide, *Arch. Environ. Health,* 28, 72, 1974.

158. **Matsushima, S.,** A statistical study on clinical cases of pesticide poisoning in rural districts, in Whither Rural Med. Proc. 4th Int. Congr. Rural Med., Usuda, 1969 (Tokyo).

159. **Kandelaki, E.,** Cases of intoxication by chlorophos and trichlormetaphos, *Gig. Primen. Toksikol. Pestits. Klin. Otravlenii,* 341, 1966, (in Russian).

160. **Zapko, V.,** Changes in some indices of the functional state of liver in cases of intoxication by chlorophos, *Gig. Primen. Toksikol. Pestits. Klin. Otravlenii,* 117, 1966, (in Russian).

161. **Ikuta, S.,** A case of intoxication by an organophosphorous pesticide, *J. Jpn. Assoc. Rural Med.,* 22, 2, 113, 1973.

162. **Babkina, I.,** Damage on nervous system at poisoning by trichlorphon, *Vrach. Delo.,* 2, 137, 1972, (in Russian).

163. **Shutov, A. and Barankina, T.,** Neurological disturbances following acute poisoning by chlorophos, *Klin. Med.,* 47, 9, 140, 1969, (in Russian).

164. **Holmes, J., Starr, H., Hanisch, R., and von Kaulla, K.,** Short-term toxicity of Mevinphos in man, *Arch. Environ. Health,* 29, 84, 1974.

165. **Stoeckel, H. and Meinecke, K.,** Über eine gewerbliche Vergiftung durch Mevinphos, *Arch. Toxicol.,* 21, 284, 1966.

166. **Chezzo, F., Perini, G., and Benetti, P.,** Sui lunomeni di sensibilizzazione da esposizione ad insecticidi organofosforis, *Ig. Mod.,* 57, 538, 1964.

167. **Bell, A., Barnes, R., and Simpson, G.,** Cases of absorption and poisoning by the pesticide Phosdrin, *Med. J. Aust.,* 1, 178, 1968.

168. **Weiner, A.,** Bronchial asthma due to organic phosphate insecticides: A case report, *Ann Allergy,* 19, 387, 1961.

169. **Perron, R. and Johnson, B.,** Insecticide poisoning, *N. Engl. J. Med.,* 281, 274, 1969.

170. **Conyers, R. and Goldsmith, J.,** A case of organophosphorus-induced psychosis, *Med. J. Aust.,* 1–58, 1, 27, 1971.

171. **Mello, D., Rodrigues Puga, F., and Benintendi, R.,** Intoxication produced by degradation products of diazinon in its use as a tick killer, *Biologica,* 38, 136, 1972.

172. **Benerjec, D.,** Pericardiris in actue diazinon poisoning. A case report, *Armed Forces Med. J. India,* 23, 187, 1967.

173. **Dochev, D. and Latifyan, K.,** Four cases of intoxication by parathion, *Savremenna Med.,* 4, 92, 1958, (in Bulgarian).

174. **Constock, E., Bickel, L., and McCormick, E.,** Acute ethion poisoning, *Texas Med. J.,* 3, 6, 1967.

175. **Hayes, A. L.,** Clinical handbook on economic poisons, U.S. Department of Health, Education and Welfare, Atlanta, 1963.

176. **Holmes, J.,** Organophosphorous insecticides in Colorado, *Arch. Environ. Health,* 9, 4, 445, 1961.

177. **Kalic-Filipovic, D., Dodic, S., Saric, S., Prodanovic, M., Arsenijevic, M., Guconic, M., and Vidakovic, A.,** Occupational health hazards in the production and application of pesticides, *Arh. Hig. Rada,* 24, 3, 333, 1973.

178. **Hayes, A. L., Wise, R. A., and Weir, T. W.,** Assessment of occupational exposure to orgasnophosphates in pest control operators, *Amer. Ind. Hyg. Assoc. J.,* 41, 1980.

179. **Franklin, C. A., Fenske, R. A., Greenhalgh, R., Mathieu, L., Denley, H. V., Leffingwell, J. T., and Spear, R. C.,** Correlation of urinary pesticide metabolite excretion with estimated dermal contact in the course of occupational exposure to guthion, *J. Toxicol. Environ. Health,* 7, 715, 1981.

180. **Wolfe, H., Durham, W. F., and Armstrong, J. F.,** Urinary excretion of insecticide metabolites, *Arch. Environ. Health,* 21, 711, 1970.

181. **Durham, W. R., Wolfe, H. R., and Elliott, J. W.,** Absorption and excretion of parathion by spraymen, *Arch. Environ. Health,* 24, 6, 381, 1972.

182. **Le Quesne, P. M. and Maxwell, I. C.,** Effect of Metriphonate on neuromuscular transmission, *Acta Pharm. Toxicol. Scand.,* 49, 99, 1981.

183. **Coye, M. J., Lowe, J. A., and Maddy, K. J.,** Biological monitoring of agricultural workers exposed to pesticides: monitoring of intact pesticides and their metabolites, *J. Occup. Med.,* 28, 8, 628, 1986.

184. **Genina, S.,** Dynamics of ChE activity in blood of aircraft-technical staff for special application at work with some organophosphorous compounds, *Materialy I Vsesojuz. Siezda Vracej Labor, I,* 1973, 62, (in Russian).

185. **Quinones, A., Bogden, D., and Nakah, E. L.,** Depressed cholinesterase activity among farm workers in New Jersey, *Sci. Total Environ.,* 6, 155, 1976.

186. **Belkin, J. and Chow, I.,** Biochemical effects of chronic low-level exposure to pesticides, *Chem. Pathol. Pharmacol.,* 9, 2, 1974.

187. **Gervais, P., Langevin, M. T., Gauthier, M., Housset, M., Boucker, M., and Mazurkievicz, S.,** Normalisation des cholinesterases chez les ouvriers desinsectiseurs par les méthodes modernes de prévention, *Arch. Mal. Prof. Med. Travai,* 35, 9, 789, 1971.

188. **Roan, C., Morgan, D., Cook, N., and Peskot, E.,** Blood cholinesterases, serum parathion concentrations and urine *p*-nitrophenol concentrations in exposed individuals, *Bull. Environ. Contam. Toxicol.,* 4, 6, 1969.

189. **Petkova, V.,** On the problem of cholinesterase activity at chronic exposure to pesticides, *Hig. i Zdraveopazvane,* 29, 1, 21, 1986, (in Bulgarian).

190. **Namba, T.,** Cholinesterase inhibition by organophosphorous compounds and its clinical effects, *Bull. OMS,* 44, 289, 1971.

191. **Namba, T., Noete, C., Jackrel, J., and Grob, D.,** Poisoning due to organophosphate insecticides, *Am. J. Med.,* 50, 475, 1971.

192. **Davies, J. E., Enos, E., Barguet, A., Morgade, C., and Danauskas, J. R.,** Pesticides monitoring studies. The epidemiological and toxicological potential of urinary metabolites, in Toxicology and Occupational Medicine, Proc. 10th Int. Ann. Conf: of Tox. and Occup. Med., 1978.

193. **Hayes, J.,** Studies on exposure during use of anticholinesterase pesticides, *Bull. WHO,* 44, 277, 1971.

194. **Ando, M., Hirosaky, S., Tamura, K., and Taya, T.,** Multiple regressive analysis of the cholinesterase activity with certain physiochemical factors, *Environ. Res.,* 33, 96, 1984.

195. **Kaloyanova, F.,** Effects of certain organophosphorous insecticides upon cholinesterase activity in persons in condition of agricultural labour, *Sb. Tr. na NIOTPZ,* 6, 105, 1959, (in Bulgarian).

196. **Kaloyanova, F.,** Cholinesterase activity as a biochemical indicator for monitoring exposure to certain pesticides, in Int. Conf. Environ. Sensing and Assessment, N.Y., Institute of Electrical and Electronic Engineers, 1975, 1.

197. **Kaloyanova, F., Benchev, I., Gheorghiev, G., Izmirova, N., and Risov, N.,** Pesticides and persistent substances. Biological specimen collection, in *Biological Specimens Used for the Assessment of Human Exposure to Environmental Pollutants,* Martinus N. Publishers, Sofia, 1979, 231.

198. Report on an international workshop, Epidemiological Toxicology of Pesticide Exposure, *Arch. Environ. Health,* 5, 25, 1972.

199. **Gage, J.,** The significance of blood cholinesterase activity measurements, *Residue Review,* 18, 159, 1967.

200. WHO Technical Reports Series, 571, Early detection of health impairment in occupational exposure to health hazards: report of a World Health Organization Study Group, 1975, 63.

201. **Vandekar, M.,** Aspects of the assessment of pesticide exposure and hazards to man, *Arch. Higienu Rada Toksikol.,* 26, 233, 1975.

202. **Miller, S. and Shah, M. A.,** Cholinesterase activity of workers exposed to organophosphorous insecticides in Pakistan and Haiti and an evaluation of the tintometer method, *J. Environ. Sci. and Health,* 17, 2, 125, 1982.

203. **Ismirova, N.,** Methods for determination of exposure of agricultural workers to organophosphorous pesticides, in *Field Worker Exposure During Pesticide Application,* Tordoir, W. F. and Hemstra-Lequin, E. A. H., Eds., Elsevier, North-Holland, Amsterdam, 1980, 169.

204. **Haskathorn, D. R., Brinkman, W. J., Hathaway, T. R., Talboth, T. D., and Tompson, L R.,** Validation of a whole blood method for cholinesterase monitoring, *Am. Ind. Hyg. Assoc. J.,* 44, 7, 547, 1983.

205. **Ellman, G. L., Courtney, K. D., Andres, V., and Featherstone, R. M.,** A new and rapid colorimetric determination of acetylcholinesterase activity, *Biochem. Pharmacol.,* 7, 88, 1961.

206. **Tordoir, W. E.,** Report of the rapporteur on: field worker exposure during pesticide application, St. in Environ. Sci., 7, Elsevier, 1980, 18.

207. **Levin, H. S., Robnitzky, R. L., and Mick, B. L.,** Anxiety associated with exposure to organophosphorous compounds, *Arch. Gen. Psychiatry,* 33, 225, 1976.

208. **Larsen, K. O. and Hanel, H. K.,** Effects of exposure to organophosphorus compounds on S-cholinesterases in workers removing poisonous depot, *Scand. J. Work Environ. Health,* 8, 222, 1982.

209. **Shafic, T. M., Bradway, D. E., Enos, H. E., and Yobs, A. R.,** Human exposure to organophosphorus pesticides. A modified procedure for the gas-liquid chromotographic analysis of alkylphosphate metabolites in urine, *J. Agric. and Food Chem.,* 21, 4, 625, 1973.

210. **Kuts, F. and Strassman, S.,** Survey of pesticide residues and their metabolites in the general population in the United States, *Ann. N.Y. Acad. Sci.,* 160, 1969.

211. **Coye, M. J., Lowe, J. R., and Maddy, K.,** Biological monitoring of agricultural workers exposed to pesticides, II. Monitoring of intact pesticides and their metabolites, *J. Occup. Med.,* 28, 628, 1986.

212. **Fournier, E., Sonnier, M., Dallys,** Detection and assay of organophosphate pesticides in human blood by gas chromatography, *Clin. Toxicol.,* 12, 4, 457, 1978.

213. **Roberts, D. V. and Trollope, I. B.,** Nerve conduction velocity and refractory period as parameters of neurutoxicity, *Electroencephalogr. Clin. Neurophysiol.,* 46, 351, 1979.

214. **Jager, K. W.,** Organophosphate exposure from industrual usage, electroneuromyography in occupational medical supervision of exposed workers, in Pesticides — induced delayed neurotoxicity, Proc. Conf., 19-20 February 1976, Washington, D.C., U.S. Environmental Protection Agency, 1976.

215. **Quantick, H. R. and Perry, I. C.,** Hazards of chemicals used in agricultural aviation; a review, Aviation, Space and Environ. Medicine, October, 581, 1981.

216. **Violante, F. S. and Roberts, D. V.,** Electromyographic study of neuromuscular function in pesticide workers, *Med. Lav.,* 3, 210, 1982.

217. **Richardson, R. J. and Dudek, B. R.,** Neurotoxic esterase, characterization and potential for a protective screen for exposure to neuropathic organophosphates, in *Proc. IUPAC Pest. Chem. Congr. Human Welfare and Environment,* Miyamota, J., Ed., Pergamon Press, 1983, 481.

218. **Bertoncin, D., Russolo, A., Caroldi, S., and Lotti, M.,** Neuropathy target esterase in human lymphocytes, *Arch. Environ. Health,* 40, 139, 1985.

219. **Lotti, M., Becker, C. E., Aminoff, M. J., Woodrow, J. E., Seiber, J. N., Talcott, R. E., and Richardson, R. J.,** Occupational exposure to the cotton defoliants DEF and merphos, *J. Occup. Med.,* 25, 517, 1983.

220. **Salas, C. and Christeva-Mircheva, B.,** Changes in some hematological indices in workers exposed to organophosphorous compounds in Algeria, *Hig. i Zdraveopazvane,* 27, 6, 541, 1984, (in Bulgarian).

221. **Maroni, M.,** Organophosphorus pesticides, in *Biological Indicators for the Assessment of Human Exposure to Industrial Chemicals,* Alessio, L., Berlin, A., Boni, M., and Roi, R., Eds., CEC, 1986.

222. **Warriner, R. A., Nies, A. S., and Hayes, W. J.,** Severe organophosphate poisoning complicated by alcohol and terpentine ingestion, *Arch. Environ. Health,* 32, 203, 1977.

223. **Hagler, J., Breachkman, R. A., Willems, J. W., Verpooten, G. A., and Debroe, M. E.,** Combined hemoperfusion-hemodialysis in organophosphate poisoning, *Appl. Toxicol.,* 1, 199, 1981.

224. **Rener, R., Robmann, K., Van Hooidonk, C., Ceulen, B. I., and Bock, J.,** Ointments for the protection against organophosphate poisoning, *Artzneim.-Forsch./Drug Res.,* 32, 6, 630, 1982.

225. WHO, Technical Report Series, 677, Recommended health-based limits in occupational exposure to pesticides, World Health Organization, Geneva, 1982.

226. NIOSH Pocket Guide to Chemical Hazards, U.S. Department of Health and Services, 1985.

227. **Kagan, Y., S.,** Principles of pesticide toxicology, UNEP/IRPTC/USSR Commission for UNEP, Centre of Inern. Projects, GKNT, Moscow, 1985.

228. **Anon.,** Pesticide residues in food, Technical Reports Series 592, World Health Organization, Geneva, 1976.

229. **Anon.,** Maximum limits for pesticide residues. Guide to Codex Recommendations concerning residues, Part 2, Codex Alimentarius Commission, Food and Agriculture Organization, Rome, 1981.

230. **Anon.,** Guideline levels for pesticide residues - Guide to Codex Recommendations concerning pesticide residues. Part 3, Codex Alimentarius Commission, Food and Agriculture Organization, Rome, 1981.

231. **Kundiev, I.,** Percutaneous absorbability of agricultural chemicals in current use, Whither Rural Medicine, in Proc. 4th Int. Congr. Rural Med., Usuda, 1969, (Tokyo), 1970, 34.

232. **Kagan, Y., Kundiev, Y., and Trotsenko, M.,** Hygiene of work during application of system organophosphorous insecticides, *Gig. Sanit.,* 6, 25, 1958, (in Russian).

Chapter 3

CARBAMATES

I. INTRODUCTION

Carbamates are part of a large group of synthetic pesticides that have developed in the last 40 years. Nowadays they are produced and used on a large scale. About 50 individual compounds are used as pesticides (Table 1).

II. PROPERTIES

Carbamates are N-substituted esters of the carbamic acid with general formula $R_1NH.CO-OR_2$.

Depending on the chemical nature of R_1, there are three classes of carbamate pesticides. If R_1 is a methyl group, the compound has insecticidal activity; if R_1 is an aromatic moiety, it acts as a herbicide, and if it is a benzimidazol moiety — as a fungicide.[1]

Carbamates are crystalline solids with low vapor pressure and low water solubility. They are moderately soluble in benzene, toluene, xylene, chloroform, and dichloromethane and lightly soluble in methanol, ethanol, acetone, and dimethyl formamide (polar organic solvents). Carbamates are formulated in granules, wettable powder, dusts, and liquids.

III. USES

Carbamates are mainly used in agriculture as insecticides, herbicides, fungicides, and nematocides. They are used for domestic desinsection and vector control in public health, as well.

IV. METABOLISM

In cases of occupational exposure carbamates are absorbed mainly through the skin and inhalation, and in small amounts — orally. The rate of dermal absorption depends on the chemical structure and the type of formulation used.[2]

Feldmann and Maibach studied percutaneous penetration and excretion of ^{14}C labeled carbaryl and baygon.[3]

After applying baygon and carbaryl to the skin for 24 h, urinary excretion on the 5th day was 19.6 and 74% respectively. These results point to a good percutaneous penetration of both compounds, especially carbaryl.

Organ distribution and excretion of carbamates are very rapid — several hours or days. They are metabolized in the human body into less toxic compounds, and in rare cases to more toxic metabolites, by a variety of chemical reactions. The general mechanism of detoxification involves the breakdown of ester bond and/or oxidation.

Aldrige and Reiner studied esterase hydrolysis of carbamates.[4] The final products are amines, CO_2, alcohol, or phenol.

A complex enzyme system is operating in the hydrolytic process.[5,6] Mixed function oxidases in the liver, kidneys and lungs are responsible for such oxydative reactions as: aromatic ring hydrolation (carbaryl), O-dealkylation (propoxur), N-methyl hydroxylation (dimethylan), N-dealkylation (aminocarb), aliphatic side chain oxidation (landrin), and thioeter oxidation (aldicarb).[7,8]

TABLE 1
General Chemical Structure and Pesticidal Activity of Carbamates[1]

Pesticidal activity	Chemical structure	Common or other names
Insecticides	or (CH₃)₂- O ‖ CH₃-NH-C-O-aryl	Aldoxycarb, allyxycarb, aminocarb, BPMC, bendiocarb, bufencarb, butacarb, carbanolate, carbaryl, carbofuran, cloethocarb, dimethalan, dioxacarb, XMC, ethiofencarb, formethanate, hoppcide, isoprocarb, trimethacarb, metolcarb, mexacarbate, pirimicarb, promacyl, promecarb, propoxur, xylycarb.
	or (CH₃)₂- O ‖ CH₃-NH-C-O-N-alkyl	Aldicarb, methomyl, oxamyl, thiofonax, thiodicarb
Herbicides	O ‖ aryl-NH-C-O-alkyl	Asulam, chlorbufam, desmedipham, phenmedipham
Fungicides	O ‖ benzimidazole-NH-C-O-alkyl	Benomyl, carbendazim, thiophanate-methyl, thiophanate-ethyl

Metabolism is often complex because of the variety of groups present in carbamates. Different metabolites are identified.

The oxidative reactions are followed by conjunction reactions to O- and N-glucoronides, sulfates, and mercapturic acid. Conjugates of carbamate hydroxy products such as glucoronides, sulfates, and mercapturic acids are rapidly excreted by urine and feces.

In a human volunteer study, carbaryl was orally administered to 2 men in gelatine capsules at a dose of 2.0 mg/kg/b.w. The overall recovery of carbaryl equivalents in urine was 26.28% (fluorometric method) or 37.8% (colorimetric method). The following metabolites were determined chromatographically in a 4-h urine sample: α-naphtyl glucoronide (10 to 15%), sulfate (6 to 11%), and 4 (methylcarbamoyloxy)-α-naphtyl glucoronide (4%). In addition, another metabolite, α-naphtylmethylimidocarbonate α-glucoronide, was identified fluorometrically.[9]

Men exposed to carbaryl dust while packing in a factory had 24-h urine specimens analyzed (the number of men, exposure period, and other conditions were unspecified). Control urines were obtained 72 h after exposure. The urinary metabolites separated by chromatography were α-naphtyl glucoronide, α-naphtyl sulfate, and unidentified neutral metabolites. The concentrations of glucoronide and sulfate in the urine were estimated to be 25 and 5 mg/l, respectively.[10]

Urine excretion of phenol derivatives after a 1.5 mg/kg oral dose of propoxur was highest in the time interval between 110 min and 285 min after dosing 177.5 and 195.6 μg/ml urine respectively. The pre-exposure level of 20 μg/ml was reached in 24 h.[11]

In a study by Dawson et al. on volunteers, 27% of the 50 mg propoxur orally administered was excreted within 8 to 10 h as α -isopropoxyphenol.[12]

Some of the carbamate metabolites in the urine are presented in Table 2.

V. TOXICITY: MECHANISM OF ACTION

Carbamates (except benzimidazol compounds) inhibit esterases. They produce carbamylation of acetylcholinesterase similar to the phosphorylation of AChE produced by organo-

TABLE 2
Metabolites From Selected Carbamates[13-17]

Carbamate Compound	Metabolites
Carbofuran	3-hydroxycarbofuran 3-ketocarbofuran 3-ketocarbofuran phenol; carbofuran phenol 3-hydroxy phenol excreted as conjugates of glucuronic acids and sulfate
Propoxur	O-hydroxyphenyl-*N*-methylcarbamate 2-iso- propoxyphenol, excreted as glucoronide
Oxamyl	Methyl-*N*-hydroxy *N'*-dimethyl-l-thiooxaminidate *N*-*N*-dimethyloxamate, excreted as glucuronide

Aldicarb	Aldicarb sulfoxide, oxime and nitrile forms, the
	  sulfone and related oxime and nitrile, as well as
	  aldehyde and acid analogous

Bendiocarb	Phenol derivatives conjugated as sulfate and betaglucuronide

TABLE 3
Reactivation Time of Carbamylated ChE by Some Carbamates

Type of carbamate	Reactivation time
	AChE (min)
Carbamyl	2–15
Methylcarbamyl	38–72
Dimethylcarbamyl	27–240
	ChE (h)
Methyl carbamyl	2.6–3
Dimethyl carbamyl	3.5–17

phosphorous compounds. Carbamates differ from OPs in the low stability of the carbamylated enzyme and the lack of the aging reaction.[18] A consequence is the rapid recovery from acute poisoning. The zone of acute effect, determined by the ratio LD_{50}:Lim ac, is much larger than that of OPs. The symptoms of intoxication develop quickly after exposure, long before the absorption of a dangerous dose, and thus prevent heavy intoxication. Recovery is very rapid, and cumulation of the effect is insignificant.

Carbamates also inhibit other esterases that have serine in their catalytic center, serine esterases or beta-esterases.[19,20]

Simeon and Reiner compared acetylcholinesterase and cholinesterase inhibition by some *N*-methyl and *NN*-dimethyl-carbamates.[21] For all tests, except one, the authors found acethyl-cholineesterase inhibited to a higher degree than plasma cholinesterase.

Carbamylated ChE is unstable.[22] The degree of inhibition depends on the rates of both the constant of inhibition and the constant of spontaneous reactivation.[23]

Table 3 shows the spontaneous reactivation (half-life) of various carbamylated cholinesterases according to Aldrige and Reiner.[4]

The longer duration of plasma ChE inhibition induced by two carbamates — propoxur and promecarb, according Plestina and Svetličić, owes to a delay in their skin resorption in the case of human occupational exposure.[24]

Some carbamates (*N*-aryl) can inhibit neuropathy target esterase (NTE), previously called neurotoxic esterase. Because no aging reaction was demonstrated with carbamates (Johnson), no delayed neuropathy effects should be expected.[25] Prior enzyme inhibition by carbamates even prevents the neurotoxic effects of several organophosphorous compounds.

The group of carbamates comprises individual compounds of widely varying acute toxicity. Oral LD_{50} in animals for this group vary from less than 1 mg/kg/b.w. to more than 5000 mg/kg/b.w. Aldicarb (temik) is the most toxic among them ($LD_{50} = 0.5$ mg/kg/b.w.). Dermal LD_{50} is over 500 mg/kg/b.w.; only aldicarb, thiofanate, methiocarb, and aminocarb are more toxic.

The inhalatory toxicity of carbamates ranges widely as well. Carbaryl produced clinical signs of intoxication in dogs at a concentration of 75 mg/m³ for 5 h (average particle size 5 µn). Guinea pigs seem to be more resistant. No clear effects were observed at 28 mg carbaryl/m³, but 350 mg/m³ produced nasal and ocular irritation and lung hemorrhages. Rats exposed to 10 mg/m³ carbaryl for 30 d did not show serious injury. LD_{50} for carbofuran is about 100 mg/m³, for methamyl — 77 mg/m³, and for aldicarb — 6.7 mg/m³.

Symptoms of intoxication appear a few minutes after carbamate is absorbed. The clinical picture results from acetylcholine accumulating at the nerve endings.

The cholinergic synapses, in which the acetylcholine is the transmitter, include the synapses of the central nervous system, neuromuscular junctions, sensory nerve endings, ganglionic synapses of both sympathetic and parasympathetic nerves, the postganglionic sympathetic nerve terminals that inervate the sweat glands and blood vessels, sympathetic nerve terminals in the adrenal medula, and all postganglionic parasympathetic nerve terminals. Acetylcholine molecules accumulated at the synapse initially cause excessive excitation and later lead to the blockage of synaptic transmission.

Extensive descriptions of the symptoms of the cholinesterase inhibition intoxication are given in *WHO TRS 356,* Namba, and Kaloyanova and Fournier.[28-30] They can be summarized as follows:

- Muscarinic effects: excessive sweating, salivation, lacrimation, bronchoconstriction, pinpoint pupils, increased bronchial secretion, abdominal cramps (vomiting and diarrhoea), and bradycardia.
- Nicotinic effects: headache, easy fatigue, dizziness, anxiety, mental confusion, convulsions, coma, and depression of respiratory center.

All these symptoms can be combined differently and vary in onset, grade, and sequence. It depends on the chemical, dose, and route of exposure. Duration of the symptoms is usually shorter than that observed in organophosphorus poisoning. In some cases the clinical picture is dominated by respiratory failure, sometimes leading to pulmonary edema.[2]

VI. EXPOSURE: DOSE-EFFECT RELATIONSHIP

In the production of carbaryl, air concentrations varied from 0.23 to 31 mg/m³.[31]

During desinsection aircraft operation, the maximum possible carbaryl exposure was calculated to be 43 mg for a person of 80 kg b.w. or 0.52 mg/kg.[32]

Exposure to carbaryl during agricultural application was studied by Leavit et al.[33] The mean dermal exposure was 128 mg/h, and the mean inhalation exposure was 0.1 mg/h. Maximally, 0.12% of the toxic dose per hour was received by the applicators.

Yakim found carbaryl concentrations in the air over treated fields ranging from 0.7 to 2.4 mg/m³.[34] During aircraft spraying of cotton fields, Yakim found 4 mg/m³ in the breathing zone of flagmen (1 mg/m³ during solution preparation) and 0.7 mg/m³ in the pilot cabin.[35]

During orchard spraying by carbaryl, Legier found air concentrations of 0.6 mg/m³ (0.18 to 0.81 mg/m³).[36] The mean respiratory exposure measured by respirator pad technique was 0.29 mg/h, and the mean dermal exposure measured by skin pads was 25.3 mg/h. The maximum total exposure was 38.83 mg/man/h, or 0.025% of the toxic dose.

The mean dermal exposure of strawberry harvesters to carbaryl was calculated at 1.45 to 2.65 mg/h; for benomyl it was at 5.39 mg/h. Exposure of blueberry harvesters to methiocarb ranged from 1.06 to 6.04 mg/h.

A rough first approximation of the dermal exposure rate for fruit harvesters, based on dislodgeable foliar residues (DFR), may be calculated using the formula proposed by Zweig in 1984.[37] Dermal exposure rate = $5 + 10^3 + DFR$ (μg/cm^2).

During urban application of carbaryl, Gold et al. calculated a maximum dermal exposure of 2.86 mg/kg/h.[38] The maximum concentration measured in the air was 0.28 mg/m^3.

VII. EFFECTS ON HUMANS

A. ACUTE INTOXICATIONS

Acute accidental and suicidal intoxications have been reported. Hayes reported two cases: a 19 month child poisoned by an unknown amount of carbaryl and a man who swallowed 250 mg of carbaryl.[39] Both developed ChE inhibition symptoms: constricted pupils, salivation, muscular incoordination (child), epigastric pain, profuse sweating, lassitude, and vomiting (man); Hayes also found inhibited blood cholinesterase. Both recovered after atropine treatment.

Farago reported a fatal case of carbaryl intoxication (0.5 l p.o.).[40] In this case it was concluded that 2-PAM (pyridine-2-aldoxime methyl chloride) application hastened the fatal outcome.

Izmirova et al. described a case of acute carbofuran intoxication.[41] A woman received a total dose of 60 mg. Slight ChE inhibition was found, but within 72 h the patient recovered completely.

Reich and Welke reported a fatal case of mexacarbate (Zectran) poisoning in which a 17-year-old boy ingested approximately 55 g (22% formulation) of mexacarbate.[42] The boy was found unconscious, with pinpoint pupils and an irregular heartbeat. He did not respond to the treatment. Bradycardia developed later with recurrent heart failure. The patient died about 4 to 4.5 h after the ingestion of mexacarbate.

Bomirska and Winiarska described two severe cases of propoxur poisoning following the ingestion of 150 to 200 ml of commercial Baygon formulation.[43] One of the patients recovered after treatment but the other died. A normal ChE suggested spontaneous reactivation of this enzyme.

Aldicarb food poisoning from contaminated melons was reported in California by local health departments and poison control centres.[44] A total of 1350 cases were reported, 692 were classified as probable on the following grounds: symptoms onset less than 2 h after consumption, melon positive for aldicarb or metabolites, multiple cholinergic symptoms occurred, and more than one person suffered ill effects from the same melon. The most severe signs and symptoms included seizures, loss of consciousness, cardial arrhythmia, hypertension, dehydration, and anaphylaxis. In all, 17 persons were hospitalized, and 6 deaths and 2 still births were reported but not confirmed by analysis for aldicarb sulfoxide.

B. VOLUNTEER STUDIES

The results of volunteer studies on propoxur performed by Wandekar et al. are summarized in Table 4.[11] As the table shows, the symptoms of intoxication appear very rapidly, 10 to 15 min after ingestion. They are related to the rate of ChE inhibition in erythrocytes but not in plasma. From the above studies, the authors concluded that a single oral dose of 0.36 mg/kg/b.w. of propoxur may be sufficient to induce initial cholinergic symptoms, while larger doses may be tolerated without symptoms if the administration is gradual.[11]

The effects of orally administered carbaryl were studied in volunteers. Daily administration of 0.06 mg/kg carbaryl for 6 weeks produced no objective evidence of health impairment. Daily administration of 0.13 mg/kg for 6 weeks produced only a slight, reversible decrease in the ability of renal proximal convoluted tubules to reabsorb aminoacids.[45] There was an increase in the ratio of urinary aminoacid nitrogen to creatinin. A single 250 mg dose of carbaryl (2 to 8 mg/kg/b.w.) resulted in moderately severe poisoning in men.[39]

In volunteers, doses of 0.7 to 1 mg baygon per kg/b.w., produced a 60% decrease in the activity of ChE.[60]

Human volunteers were given single oral doses of aqueous aldicarb solution at doses of 0.025, 0.05, and 0.1 mg/kg/b.w. Mild depression of whole blood ChE was noted in all cases. Persons dosed by 0.1 mg/kg presented cholinergic symptoms such as nausea, sweating, pinpoint pupils, and salivation.[26]

C. OCCUPATIONAL SHORT- AND LONG-TERM EXPOSURE

There is no sufficient information on the short- and long-term effects of occupational exposure to carbamate pesticides.

Some reports suggest that, during carbamate production and application in agriculture and public health purposes, there are possibilities for adverse health effects, even for heavy intoxications.[47-50]

However, heavy occupational intoxications by carbamates are very rare. Tobin underlines two main reasons for the absence of heavy intoxications: (1) the very short time between exposure and onset of symptoms — half an hour or less and (2) the lack of symptom progression because of the high ratio between a median effective dose and a lethal dose of carbamates.[48] Appearance of slight symptoms may be expected long before a dangerous dose is absorbed.

1. Short-Term Exposure

Richardson and Battelse reported an acute accidental occupational intoxication by Zectran (merbate).[49] A co-pilot of a spray aircraft was exposed to fine aerosol spray for 110 min; the pump failed and there was a pinhole leak into the fuselage. Symptoms of moderate intoxication were observed. ChE activity in the blood was severely depressed, but it returned to normal after 3 d.

Tobin reported five case histories of acute occupational intoxication by carbofuran.[48] Air concentrations in a formulation factory were 0.027 to 0.642 mg/m^3 (after filtration of a nonrespirable fraction — 0.005 to 0.012 mg/m^3). Dermal absorption of carbofuran was possible as well. The following symptoms were observed:

Case I	Case II	Case III	Case IV and V
Weakness	Weakness	Poor coordination	Burning eyes
Profuse perspiration	Perspiration	Loss of depth perception	Dimmed vision
Upset stomach	Nausea	Nausea	Blurred vision
Blurred vision	Blurred vision	Weakness	Local effects
No treatment	Atropin treatment	No treatment	

All persons recovered after several hours.

The author compared these data with those of a monkey experiment. The minimum air concentration to depress ChE in monkeys was 0.57 mg/m^3; the maximum nondepression level was 0.37 mg/m^3. One animal developed tremors at 1.3 mg/m^3 after 30 min exposure, and another — emesis after 12 min exposure to 0.86 mg/m^3. As a threshold limit the author suggests 0.25 mg/m^3.

Reportedly many occupational intoxications are provoked by aldicarb. The effects of aldicarb application in a banana plantation workers were studied. From a total of 15 applicators, two showed a greater than 25% reduction of blood ChEA, 29 and 50% respectively. During the second day of the study, 6 workers had ChE below the normal range; 1 worker presented symptoms of nausea, stomach aches, and headaches.[26]

During malaria control operations in Iran, using the insecticide propoxur, Vandekar et al. performed clinical and biochemical studies.[50] Minor symptoms related to overexposure were recorded among 24 operators during 6 weeks: 20 headache cases, 11 nausea, 2 giddiness, 2 blurred vision, 1 weakness, 7 increased sweating, 2 vomiting, and 1 pinpoint pupil. Vandekar et al. found a pronounced fall in ChEA, to 22% of the normal in some cases, during the work and rapid recovery after discontinuation of exposure. Erythrocyte ChE proved to be a more sensitive test than plasma ChE. No cumulative inhibitory effect was shown.

Vandekar et al. also studied the general population during this operation. Out of 11,000 inhabitants, 69 complained of mild cholinergic symptoms when, contrary to instructions, they entered their houses. The symptoms were headaches, nausea, sweating, and vomiting. They subsided in 2 to 3 h.

Vandekar evaluated propoxur in a Nigerian village-scale trail, as well.[51] Propoxur application provoked mild to moderate intoxication in all spraymen, in spite of their protective clothing and masks. Most of the symptoms disappeared gradually in a few hours. Beside the usual symptoms, pressure inside the chest was the more persistent and outstanding complaint. Marked depression of plasma ChEA was found in all spraymen present on the following day. Three days later, the effects vanished. Vandekar found a marked increase of phenyl derivatives during the afternoon of the spraying. In villagers, phenol excretion increased from 21.0 mg/ml before spraying to 51.7 on the next day. This effect lasted for several weeks.

While applying Baygon (propoxur) as a malaria insecticide in El Salvador, 4 of the 6 spraymen experienced mild anticholinesterase symptoms.[52]

In 1962 the Union Carbide Corporation Medical Department reported a carbaryl accident involving 14 employees. On the actual day of the accident 7 had worked 8 to 16 h and complained of nausea, dizziness, and headache; they attributed it to carbaryl dust and vaporized carbaryl inhalation. After 1 d, conjugated a-naphtol and ChEA were determined. The mean conjugated a-naphtol in the urine of the 7 men was 14.2 ml/l (in a 2 to 32 mg/l range). No environmental data were reported, but the verbal descriptions given by the employees suggested concentrations well above those normally encountered.[53]

In a short-term inhalation-absorption study, 2 workers who wore either respiratory or skin protection equipment, but not both, were exposed to approximately 50 mg/m^3 airborne carbaryl during 2 work days. No apparent adverse effects were observed. The urinary a-naphtol was 36 to 90 mg/l on the first day and 23 to 24 mg/l on the second day. No noticeable effects on either employee were observed or expressed.[53]

In an urine specimen collected 18 h after an acute intoxication by 250 mg carbaryl, a level of 31.4 ppm l-naphtol was found.[39]

Vandekar also evaluated carbaryl in a village-scale trial in Nigeria.[51] The risk for the population exposed to carbaryl was assessed. One sprayman had a pronounced rash on his back after the sprayer splashed him. Slight depression of plasma ChEA was found in all spraymen the day after spraying. Determination of l-naphtol derivatives in urine did not show any increase on the first and second day. A slight enhancement was observed on the sixth day. The excretion of l-naphtol in the urine of villagers varied from 24 to 36.7 µg/ml before spraying and from 40.15 to 60 µg/ml after spraying.

Leavit et al. studied carbaryl exposure during agricultural application.[33] The mean dermal exposure was 128 mg/h, and the mean inhalation exposure was 0.1 mg/h. The maximum level of toxic dose received by the applicators was 0.12%/h. When the total exposure was less than 0.02% of the toxic dose, no symptoms of ChEA inhibition were reported.

Gold et al. studied the urban application of carbaryl.[38] The maximum dermal exposure recorded in this study was 2.86 µg/kg/h. The maximum air concentration was 0.28 mg/m^3. The decrease of ChEA in serum and erythrocyte was insignificant. The mean total carbaryl exposure of the applicator, expressed as a percentage of toxic dose per hour, was 0.01% with a maximum of 0.08%. This exposure rate is below that which should be considered a risk to urban applicators. The authors reported no symptoms of intoxication.

In a study of 19 agricultural workers by Yakim, whole blood ChEA was measured for 3 to 4 d, before and after 4 to 6 h exposures to airborn carbaryl.[54] Men exposed to a mean airborne carbaryl concentration of 2 mg/m^3 ChEA showed a 13 to 30% decrease. No objective signs of ill health were observed. In the same study, a mean carbaryl concentration of 0.7 mg/m^3 was reported in the cabin of an airplane engaged in aerial carbaryl application, but no changes were reported in the biological parameters of the pilots.

Skalsky et al. demonstrated that in rats carbaryl metabolites are present in blood at similar concentrations as saliva.[55] A correlation was found between the ChEA, blood, and saliva concentrations of carbaryl metabolites.

For carbamate herbicides (alkylesters of alkylcarbamic acid), especially phenylcarbamates, the mechanism of action is related to methemoglobinenia and hemopoietic disorders. Stimulation of erythropoiesis in bone marrow leads to erythrocytosis. Decrease of erythrocytes, leucopenia, thrombopenia, and hemolytic anemia may occur. Toxic granules in neutrophils may also be found.

Certain carbamates can irritate the skin, eyes, and respiratory tract. Tobin described two cases of direct effect on the eyes (irritation and blurred vision) during carbofuran exposure.[48] These two persons did not develop symptoms of systemic poisoning.

Vandekar reported a skin rash on a sprayman accidentally splashed with a carbaryl formulation.[51] Although carbamate compounds have not generally been implicated as a cause of dermatitis or allergic skin reactions, the author suggests that certain individuals can develop this after unusually heavy exposure.

2. Long-Term Exposure

There are only a few publications concerned with chronic and long-term effects of carbamates.

Workers exposed to carbaryl dust, sometimes at concentrations as high as 40 mg/m^3 under abnormal conditions but usually lower, showed a slight depression of blood cholinesterase but not intoxication. Best and Murray[31] found a 0.23 mg/m^3 concentration of sevin in the air of less contaminated areas of a manufacturing unit, and 29 mg/m^3 in the bagging area during abnormal conditions. ChEA levels were within normal values, or only occasionally, slightly depressed.

Humans, exposed to sevin excrete *l*-naphtol conjugated either as sulphate or glucoronide. Elevated levels of *l*-naphtol were found in 41% of the samples ($1000 \text{ µg}/100 \text{ ml}$ or 10 ppm total *l*-naphtol compared with 150 to $400 \text{ µg}/100 \text{ ml}$ urine or 1.5 to 4 ppm in the control group).[31] No symptoms were recorded. In an acute intoxication case 3140 µg *l*-naphtol per 100 ml urine was found.

Shafic et al. found 6.2 to 78.8 ppm of *l*-naphtol in urine samples obtained from formulators, and 0.07 to 1.7 ppm in agricultural workers.[56]

In chronic carbaryl intoxication the following symptoms are reported: disfunction of the thyroid gland, hyperactivity of the suprarenals, depression of the reproductive gland function, and allergic reactions of the skin and respiratory system.[57]

Morse and Baker studied workers in a methomyl production plant.[58] Subjective and objective symptoms were recorded often: for a 2-year period, 2.55 symptoms or illness episodes per worker. Symptoms such as headache, giddiness, pupillary constriction, blurred vision, weakness, cramps, diarrhea, and chest discomfort are, according to the authors, more indicative of carbamate exposure than cholinesterase determination because of the rapidly reversible ChEA inhibition.

Men having worked from 1 to 18 years in a carbaryl production plant showed a higher rate of sperm shape abnormalities. No dose dependance was established, but men having worked less than 6 years were most affected. Some results suggest that the effects may be nonreversible.[59]

Chromosome studies on the blood cultures of 20 workers engaged in benomyl production showed no increased frequency of structural chromosome aberrations.[60] Regan et al. reported irreversible DNA binding in human skin cells that were treated *in vitro* by nitrosocarbaryl.[61]

TABLE 4
Relationship Between Dose and Effect

Dose-Propoxur	Biological indicator of exposure	Symptoms after ingestion
1.5 mg/kg/b.w. single/oral	ChEA inhibition to 27% of the normal in 15 min after ingestion	15 min: moderate discomfort, pressure in the head 18 min: blurred vision 20 min: sweating, pale face, tachycardia, moderate hypertension 30 min: vomiting 2 h recovery
	45th min. Recovery to 55% of normal 2–3 h — recovery to normal Excretion of phenol derivatives by urine (45% of the ingested) 1 h 50 min 177.5 µg/ml 4.45 h — 195.6 µg/ml 24 h reach normal values 20 µg/ml	
0.36 mg/kg/b.w. single/oral	Inhibition of Er. ChEA to 57% of normal after 10 min 3 h recovery of ChEA to the normal	10–15 min stomach discomfort, blurred vision, moderate facial redness, sweating
5 doses × 0.15 or × 0.20 mg/kg/b.w. at 30 min interval	Inhibition of Er ChEA to 60% of normal level	No symptoms

D. BIOLOGICAL AND HEALTH INDICATORS OF EXPOSURE

As an early indication of occupational overexposure, most of the authors recommend minor complaints (well differentiated by other illness symptoms) such as headaches, nausea, stomach pain, sweating and vomiting, tiredness, and blurred vision. ChEA determination, its daily fluctuation and rapid reactivity are, nevertheless, a useful tool for biologically monitoring the effects of exposure, if pre-exposure levels are known.[62]

Erythrocyte and plasma cholinesterase are used to monitor exposure and systemic absorption of carbamate insecticides.[20] Carbamates are direct inhibitors of acetylcholinesterase (acetylcholine acetyl-hydrolase 3.1.1.7), which is present in human erythrocytes, nerves, synapses, and skeletal muscles; they also directly inhibit cholinesterase (acetylcholine acetyl-hydrolase 3.1.1.3), which is present in human plasma (serum) and liver. These esterases are differentiated by substrate specificity (acetylcholinesterase hydrolyses acetyl-β-methylcholine but very little benzoylcholine, propionylcholine, or butyrylcholine; the opposite is true for cholinesterase).

Inhibition of erythrocyte cholinesterase activity correlates with symptoms of intoxication (Table 4).[11]

During Baygon application to control malaria, the blood ChEA in the sprayers and population was determined using the Lovibond comparator (Tintometer method). A relationship between the reduction of ChEA to 62.5% of normal and subjective symptoms such as nausea, vomiting, tachycardia, and abdominal pains were reported. In another case, 50% inhibition was related with vertigo, vomiting, headaches and severe perspiration after 6 d contact.[63] Neither clinical nor subjective symptoms were related to urinary excretion of *l*-naphtol in excess of 1000 µg/100 ml.[31]

Plasma ChE is synthesized in the liver, and its activity is a sensitive indicator of liver function.

As the normal value for either enzymes varies widely (up to 15 to 25% for ErChE), the determination of pre-exposure values is an important prerequisite for the later assessment of the degree of depression.[64] It is the proportionate depression from the normal activity in the individual that is important and not a single numerical value.

The whole blood ChEA inhibition reduced by 30% of the pre-exposure level was proposed as a hazard level.[65]

Later, Gage proposed the same degree of ChEA inhibition for both plasma and erythrocytes.[66] The sensitivity of the method used is of great importance.

The WHO Working Group suggested the following are the hazard levels for cholinesterase depression: acetylcholinesterase activity reduced by 30% of the pre-exposure value requires repeated determination after appropriate intervals and appraisal of the general and individual work situations and (2) acetylcholinesterase activity reduced by 50% of the pre-exposure value requires immediate action, including temporary removal from further exposure and appraisal of the work situation.[67]

ChE activity in whole blood, plasma, or red cells inhibited by less than 25 to 30% was considered to be physiologically unimportant.[53] Nackathorn et al. reported on the validation of a whole blood method for cholinesterase monitoring; it was developed by one of the authors.[68] Whole blood cholinesterase activity is measured spectrophotometrically from a 50 µl blood sample taken from the earlobe. The advantage is that the earlobe is less contaminated than the finger.

They also underline the consideration that the relative change from the normal values of an individual should be the most significant parameter for evaluating exposure.

The test data showed that more than 97% of the activity values will be 80% of the normal. The coefficient of variation for 12 groups studied ranged from 9.7 to 13% due to analysis, sample collection, and personal variation. The authors concluded that when the ChE value reaches 70% of the normal, it is almost certain the depression is due to exposure to a cholinesterase inhibitor, as suggested earlier by the field method applied by Kaloyanova.[65] The following proposals were made by the authors: (1) baseline values should be established by 3 separate tests prior to exposure, (2) at 70% of the baseline values, investigate and correct possible exposure situations, and (3) at 60% of baseline value, remove the employee from exposure to anticholinesterase agents. Carbamylated ChE spontaneously reactivates, its half-life being 3 to 3.5 h.[22] In field conditions, the opposite can happen. During a series of operational field trials with propoxur, it was observed that the inhibition of the whole blood cholinesterase was greater when the samples were stored before the assay, due to continuous unknown fluctuations. Both these reasons limit the ChE biological monitoring for carbamates. Wilhelm et al. compared the methods for measuring ChE inhibition *in vitro* by 4 carbamates.[65] When plasma ChE was measured by spectrophotometric methods and Acholest, no difference was found. The whole blood activity was slightly higher when measured by tintometric method. The explanation of this phenomenon, according to the authors, was that when using tintometric method, blood samples are diluted 100-fold; this prevents further enzyme inhibition during the assay. The authors also noted that when undiluted plasma was used, and if inhibition was not at a steady rate, further inhibition of ChE might occur during the assay.

E. METHODS AVAILABLE FOR ChEA DETERMINATION

Elman et al.[69] developed a photometric method for determining AChEA. It is based on the increase in yellow color produced by thiocholine when it reacts with dithiobisnitrobenzoate ion. This is the method mostly used nowadays for research and other purposes.

A radiometric method for estimating blood cholinesterase in field conditions was proposed by Wintergham and Disney.[70]

Izmirova reported for the advantages of the Paper test in field conditions.[71]

For storing whole blood samples inhibited by carbamates a 300-fold dilution with a pH 5.0 buffer at 4°C and measurement of ChE activity within 7 h after dilution was suggested.[72,73]

Metabolites of carbamates in urine are used as a biological monitoring tool. Carbaryl is

TABLE 5
Maximum Permissible Levels for Carbaryl

TWA	Short-term	Country
1 mg/m^3	—	Bulgaria, Hungary, U.S.S.R.
2 mg/m^3	5 mg/m^3	Romania
5 mg/m^3	—	Argentina, Austria, Belgium, Fed. Rep. Germany, Netherlands, Switzerland, U.S., Yugoslavia
	10 mg/m^3	Argentina, U.S.

excreted as *l*-naphtol in urine. A rapid and sensitive method for determining *l*-naphtol in human urine was developed by Shafik et al. using electron capture gas chromatography detection.[56] The results are reproducible over a wide range of concentrations. The limit of detectability has been determined as 0.02 ppm (0.02 mg/l).

Generally colorimetric determination of free and conjugated *l*-naphtol concentration in urine is used.[31]

Urine excretion of *l*-naphtol significantly above 1000 mg/100 ml urine indicates absorption and metabolism of sevin in an exposed person.[31]

L-naphtol levels in urine can be used as a biological indicator to detect carbaryl absorption (in the absence of *l*-naphtol in the working environment). However, owing to the lack of data on the relationship between *l*-naphtol levels in urine and the effect of exposure, no health-based biological limits can be proposed at present.[53]

Determination of phenol excretion in urine after propoxur exposure is also used as a biological indicator.[12] Gas chromatography and simpler colorimetric methods for determining 2-isopropoxyphenol are described.

F. TREATMENT OF CARBAMATE INTOXICATIONS

Cases of carbamate intoxication should be treated as an emergency. There are three main principles: to minimize the absorption, to give general supportive treatment, and to give specific pharmacological treatment.[1]

Alkaline soap and sodium bicarbonate are used for skin decontamination. Extensive eye irrigation with water or saline should be performed. Gastric lavage with bicarbonate solution and charcoal in unconscious patients or forced vomiting in conscious ones are recommended. Artificial respiration via a tracheal tube should be started at the first sign of respiratory failure.[1]

As a specific pharmacological treatment, atropin should be given, beginning with 2 mg i.v. and repeated at 15 to 30 min intervals to maintain fully atropinized.

Precaution should be taken not to overdose atropin. Oxime reactivators are not recommended; they may aggravate the health situation. Diazepam should be included in the therapy (10 mg s.c. or i.v.) to relieve anxiety, when necessary, and counteract some CNS derived symptoms, which are not affected by atropin.

VIII. PREVENTION

The permissible levels of some carbamates in different environmental media are summarized below and in Tables 5 and 6.[1,74,75]

Immediately dangerous to life or health (IDLH) is a concentration of 600 mg/m^3 carbaryl in the air.

In the U.S.S.R. MPL for carbaryl in ambient air is 0.02 mg/m^3 for short-term (30 min), and 0.01 mg/m^3 for 24-h exposure; MPL in drinking water is 0.1 mg/l.

TABLE 6
Maximum Permissible Levels for Carbamates

Carbamate	U.S.S.R. mg/m³	U.S. TWA ppm	U.S. TWA mg/m³	U.S. STEL ppm	U.S. STEL mg/m³
Acylate [acetoxyisopropyl plenylcarbamate]	2.0				
Benomyl		0.8	10	1.3	15
Betonal (Fendifam) [3-methoxycarbonyl aminophenyl-*N*- (3-methylphenyl-)carbamate]	0.5				
Carbin [4-chlorobutin-2yl -*N*-(3-clorophenyl)]	0.5				
Carbofuran		0.1			
Chlor IPC [isopropyl-*N*-(3-chloro-phenyl) carbamate]	2.0				
Eptam [ethyl-*N*, *N*-di- propylthiocarbamate]	2.0				
IPC [isopropyl-*N*-phenyl-carbamate]	2.0				

The acceptable daily intakes for individual carbamates

aminocarb	0–0.004	mg/kg/b.w.
bendiocarb	0–0.02	mg/kg/b.w.
benomyl	0–0.02	mg/kg/b.w.
carbedazim	0–0.01	mg/kg/b.w.
carbofuran	0.01	mg/kg/b.w.
ethiofencarb	0.1	mg/kg/b.w.
methiocarb	0.001	mg/kg/b.w.
oxamyl	0.03	mg/kg/b.w.
pirimicarb	0.02	mg/kg/b.w.
thiophanate-methyl	0.08	mg/kg/b.w.
carbaryl	0.01	mg/kg/b.w.
propoxur	0.02	mg/kg/b.w.
aldicarb	0.005	mg/kg/b.w.

Acceptable residues of carbaryl in food

Czechoslovakia	0.2–10	mg/kg
Germany	0.5–5	mg/kg
EEC	1–3	mg/kg
FAO/WHO	milk 0.1 mg/kg; meat 0.2 mg/kg	
USSR	0.1–1	mg/kg

Acceptable residues of other carbamates:

pyrimor	0.5–0.1	mg/kg in beans
propoxur	0.5	mg/kg in potatoes, carrots, 0.05 mg/kg in meat, milk
betanal	0.03	mg/kg in sugar beet

Carbamates are easily degradable in the environment, including vegetables and fruits. That is why they do not create residue problems.

IX. CONCLUSION

The group of carbamate pesticides includes about 50 individual compounds, used as insecticides, herbicides, fungicides, and nematocides.

They penetrate easily through the skin. Acute toxicity differs for individual compounds varying from 0.5 mg to 5000 mg/kg/b.w. In the organism carbamates are subjected to hydrolytic, oxidative, and conjunctive reactions, and they are excreted through the urine. Carbaryl is excreted as I-naphtyl glucoronide, I-naphtyl sulfate, etc. Carbamates do not accumulate in the tissues.

Carbamates (except benzimidazol compounds) are cholinesterase inhibitors. Carbamylation of cholinesterase is reversible; no aging reaction occurs. For this reason, recovery from acute intoxication is rapid.

Symptoms of intoxication are demonstrated within minutes after absorption. The symptoms are based on acetylcholine accumulation, and the clinical picture is similar to that of intoxication with organophosphorous compounds.

Acute intoxications with carbamates are the main health problem; they may have a lethal issue. Chronic intoxications are not likely to occur. Data on long term effects are controversial and almost exclusively concern animal experiments. Only in one study, in a carbaryl production plant, morphological sperm abnormalities were found in some workers. Nitrozation of carbaryl leads to the formation of nitrosocarbaryl, which may present carcinogenic risk.

REFERENCES

1. Carbamate pesticide, a general introduction. Environmental Health Criteria 64, World Health Organization, Geneva, 1986.
2. FAO/WHO, Pesticide residues in food, FAO Plant Production and Protection Paper 56, (Report of the 1983 Joint Meeting of FAO Panel of Experts on Pesticide Residues in Food and the Environment and the WHO Expert Group on Pesticide Residues), Food and Agriculture Organization, Rome, 1984.
3. **Feldaman, R. J. and Maibach, H. I.,** Percutaneous penetration of some pesticides and herbicides in man, *Toxicol. Appl. Pharmacol.,* 28, 126, 1974.
4. **Aldridge, W. N. and Reiner, E.,** *Enzyme Inhibitors as Substrates. Interaction of Esterases with Esters of Organophosphorous and Carbamic Acids,* Elsevier, Amsterdam, 328, 1982.
5. **Reiner, E. and Skrinjaric'Spoljar, M.,** Hydrolysis of some monomethylcarbamates in human sera, *Croat. Chem. Acta,* 48, 87, 1968.
6. **Sakai, K. and Matsumura, F.,** Degradation of certain organophosphate and carbamate insecticides by human brain esterases, *Toxicol. Appl. Pharmacol.,* 19, 660, 1971.
7. **Fukuto, T. R.,** Metabolism of carbamate insecticides, *Drug Metab. Rev.,* 1, 117, 1972.
8. **Oonithan, E. S. and Casida, J. E.,** Oxidation of methyl and dimethylcarbamate insecticide chemicals by microsomal enzymes and anticholinesterase activity of the metabolites, *J. Agr. Food Chem.,* 16, 28, 1968.
9. **Knaak, J. B., Tallant, M. J., Kozbelt, S. T., and Sullivan, L. J.,** Metabolism of carbaryl in man, monkey, pig and sheep, *J. Agr. Food Chem.,* 16, 1968, 465.
10. **Knaak, J. B., Tallant, M. J., Bartley, M. Y., and Sullivan, L. J.,** The metabolism of carbaryl in the rat, guinea pig and man, *J. Agr. Food Chem.,* 13, 1965, 537.
11. **Vandekar, M., Plestina, R., and Wilhelm, K.,** Toxicity of carbamates for mammals, WHO Bull. No 44, World Health Organization, Geneva, 1971, 241.
12. **Dawsen, I. A., Heath, D. F., Rose, I. A., Thain, E. M., and Ward, I. B.,** The excretion by humans of the phenol derived *in vivo* from 2-iso-propoxyphenyl-*N*-methylcarbamate, WHO Bull. No 30, World Health Organization, Geneva, 1964, 127.
13. Data Sheet on Pesticides No. 25, Propoxur, World Health Organization/Food and Agriculture Organization.
14. Data Sheet on Pesticides No. 52, Bendiocarb, World Health Organization/Food and Agriculture Organization, 1982.
15. Data Sheet on Pesticides No. 53, Aldicarb, World Health Organization/Food and Agriculture Organization, 1982.

16. Data Sheet on Pesticides No. 54, Oxamyl, World Health Organization/Food and Agriculture Organization, 1983.
17. Data Sheet on Pesticides No. 56, Carbofuran, World Health Organization/Food and Agriculture Organization, 1986.
18. **Aldridge, W. N.**, Chemical structure and toxicity of pesticides, in Toxicology of Pesticides, Interim document No. 9, World Health Organization/Association of European Operational Research Societies, Copenhagen, 21, 1982.
19. **Sakai, K. and Matsumura, F.**, Esterases of mouse brain active in hydrolysing organophosphate and carbamate insecticides. *J. Agr. Food Chem.*, 16, 803, 1968.
20. **Aldridge, W. H. and Magos, L.**, Carbamates, thiocarbamates and dithiocarbamates, Commission of the European Communities, Luxembourg, 1978.
21. **Simeon, V. and Reiner, E.**, Comparison between inhibition of acetylcholinesterase and cholinesterase by some *N*-methyl and *NN*-dimethyl carbamates, *Arch. Hig. Rada Toxicol.*, 24, 199, 1973.
22. **Reiner, E.**, Spontaneous reactivation of phosphorylated and carbamylated cholinesterases, *Bull. Org. Mond. Santé*, 44, 109, 1971.
23. **Reiner, E. and Aldridge, W. H.**, Effect of pH on inhibition and spontaneous reactivation of acethylcholinesterase treated with esters of phosphorus acids and carbamic acids, *Biochem. J.*, 105, 171, 1967.
24. **Plestina, R. and Svetlicic**, Toxic effects of two carbamate insecticides in dogs, *Arch. Hig. Rada*, 24, 217, 1973.
25. **Johnson, M. R.**, Organophosphorus and other inhibitors of brain "neurotoxic esterase" and the development of delayed neurotoxicity on brain, *Biochem. J.*, 120, 53, 1970.
26. FAO/WHO, Pesticide residues in food, FAO Plant Production and Protection Paper No. 37, (Report on the 1981 Joing Meeting of the FAO Panel of Experts on Pesticide Residues in Food and the Environment, and WHO Expert Group on Pesticide Residues), Food and Agriculture Organization, Rome, 1982.
27. Data Sheet on Pesticides No. 55, Methomyl, World Health Organization/Food and Agriculture Organization, 1982.
28. WHO Technical Report Series No. 356, Safe use of pesticides in public health, World Health Organization, 1967.
29. **Namba, T., Nolte, C. I., Jackrel, G., and Grob, D.**, Poisoning due to organophosphate insecticides: acute and chronic manifestations, *Am. J. Med.*, 50, 475, 1971.
30. **Kaloyanova, F. and Fournier, E.**, Les Pesticides de l'homme, *Coll. de méd. legale et toxicol médicale*, Masson & Cie, Paris, 1971, 166.
31. **Best, E. and Murrey, B.**, Observations on workers exposed to sevin insecticide, a preliminary report, *J. Occup. Med.*, 4, 10, 507, 1962.
32. **Cameron, E. A., Loerch, C. R., and Mamma, R. O.**, Incidental and indirect exposure to three chemical insecticides used for control of gipsy moth Lymantria dispar (2). *Z. Angew. Entomol.*, 99, 241, 1988.
33. **Leavit, J. R. C., Gold, R. E., Holcslaw, T., and Tupy, D.**, Exposure of professional pesticide applicators to carbaryl, *Arch. Environ. Contam. Toxicol.*, 11, 57, 1982.
34. **Yakim, V. S.**, Inhalatory toxicity of sevin. *Mater. Konf. Gig. Sanit. Vracej Moldavskoj SSR*, p.115, 1965, (in Russian).
35. **Yakim, V. S.**, Data for substantiating the MAC values for sevin in the air of the working zone, *Gig. Sanit.*, 4, 23, 19, (in Russian).
36. **Jegier, Z.**, Health hazards in insecticide spraying of crops, *Arch. Environ. Health*, 8, 1964.
37. **Zweig, G., Leffingwell, I., and Popendorf, W.**, The relationship between dermal pesticide exposure by fruit harvesters and dislodgeable foliar residues, *S. Environ. Sci. Health*, 20, 27, 1985.
38. **Gold, R. E., Leavitt, J. R. C., Holcslaw, T., and Tupy, D.**, Exposure of urban applications to carbaryl, *Arch. Environ. Contam. Toxicol.*, 11
39. **Hayes,** *Clinical Handbook on Economic Poisons*, U.S. Department Health, Education and Welfare, Atlanta, 1963.
40. **Farago, A.**, Fatal suicidal poisoning with sevin (*l*-naphthyl-*N*-methyl carbamate), *Arch. Toxicol.*, 24, 309, 1969, (in German).
41. **Izmirova, N., Milceva, V., Monov, A., and Kaloyanova, F.**, Acute carbofuran intoxication, *Hig. Zdraveop.*, 24, 445, 1981, (in Bulgarian with English summary).
42. **Reich, G. H. and Welke, J. O.**, Death due to a pesticide, *N. Engl. J. Med.*, 247, 1432, 1966.
43. **Bomipska, T. and Winiarska, A.**, Toxicology of carbamate in the light of intoxication with Baygon insecticide, *Pol. Tyg. Lek.*, 27, 1448, 1972, (in Polish).
44. **Jackson, R. J., Statton, S. W., Goldman, L. K., Smith, D. F., Pond, E. M., Epstein, D., Neutre, R. R., Kelter, A., and Kizer, K. W.**, Aldicarb food poisoning from contaminated melons, California, *MMWR*, 35, 16, 258, 1986.
45. **Wills, J. H., Jameson, E., and Coulsion, F.**, Effects of oral doses of carbaryl on man, *Clin. Toxicol.*, 1, 3, 265, 1968.
46. **Medved, L. I.**, Toxicology of pesticides, in *Occupational Health in Agriculture*, Medved, L. I. and Kundiev, J. I., Eds., *Medicina*, Moscow, 1981, 156, (in Russian).

47. **Comer, S. W., Staff, D. C., Armstrong, I. F., and Wolf, H. R.,** Exposure of workers to carbaryl, *Bull. Env. Contam. Toxicol.,* 13, 385, 1975.
48. **Tobin, J.,** Carbofuran. A new carbamate insecticide, *J. Occ. Med.,* 12, 1, 16, 1970.
49. **Richardson, E. M. and Battese, R. I., Jr.,** An incident of Zectran, *J. Maine Med. Assoc.,* 64, 158, 1973.
50. **Vandekar, M., Hedayat, S., Plestina, R., and Ahmady, G.,** A study of the safety of O-isopropoxyphenol methylcarbamate in an operational field trial in Iran, WHO Bull. No. 38, World Health Organization, Geneva, 1968.
51. **Vandekar, M.,** Observations on the toxicity of carbaryl, folithion and 3-isopropylphenyl *N*-methylcarbamate in a village scale trial in Southern Nigeria, WHO Bull. No. 33, World Health Organization, Geneva, 1965, 107.
52. **Quinby, G. E. and Babione, R. W.,** Toxicological observations on the use of baygon (2-isopropoxyphenyl *N*-, ethylcarbamate) as a malaria control insecticide in El Salvador, *Ind. Med. Surg.,* 35, 596, 1966.
53. Recommended health based limits in occupational exposure to pesticides, WHO Technical Report Series No. 677, World Health Organization, Geneva, 1981.
54. **Yakim, V. S.,** Data for substantiating the maximum permissible concentration of sevin in the air of working zones, *Gig. Sanit.,* 32, 29, 1967, (in Russian).
55. **Skalski, H. L., Lane, R. W., and Borzellece, J.,** Excretion of carbaryl into saliva of the rat and its effects on cholinesterase, in *Toxicology and Occupational Health,* Deichman, W., Ed., Int. Conf. on Toxicol. and Med., 10-11 May 1978, Elsevier 1979, 349.
56. **Shafic, M. T., Sullivan, H. C., and Enos, H. F.,** A method for determination of 1-naphthol in urine, *Bull. Environ. Contam. Toxicol.,* 6, 34, 1971.
57. **Krasnjuk, E. M. and Lubjanova, I. P.,** *Occupational Diseases in Agricultural Workers,* Kundiev, J. K. and Krasnjuk, E. P., Eds., Zdorovje, Kiev, 1983.
58. **Morse, D. L. and Baker, E. I.,** Propanyl, chlordane and methamyl toxicity in a pesticides manufacturing plant, *Clin. Toxicol.,* 15, 13, 1979.
59. **Wyrobek, A., Watchmaker, G., Gordon, L., Wong, K., Moore, D., and Whorton, D.,** Sperm shape abnormalities in carbaryl exposed employers, *Environ. Health Perspect.,* 40, 255, 1981.
60. **Ruzicska, P., Peter, S., and Czeizel, A.,** Studies on the chromosomal mutagenic effect of benomyl in rat and humans, *Mutat. Res.,* 29, 201, 1975.
61. **Regan, J. D., Setlow, R. B., Francis, A. A., and Lyimsky, W.,** Nitrosocarbaryl: its effect on human DNA, *Mutat. Res.,* 38, 293, 1979.
62. **Kaloyanova, F.,** Cholinesterase activity as an indicator for monitoring exposure to certain pesticides, Int. Conf. Environ. Sensing and Assessment, I, Nevada, 13-2, 1-4, 1975.
63. **Montazemri, R.,** Toxicological studies on baygon insecticides in Shabarkarah, *Acta Iran. Trop. Geogr. Med.,* 21, 186, 1969.
64. **Augustinsson, V.,** The normal variation of human blood cholinesterase activity, *Acta Physiol. Scand.,* 35, 40, 1955.
65. **Kaloyanova, F.,** Effects of certain organophosphorous insecticides upon cholinesterase activity in persons in conditions of agricultural labour, *Sb. Trudove NIHPZ,* 6, 105, 1959.
66. **Gage, J.,** The significance of blood cholinesterase activity measurement, *Res. Rev.,* 18, 159, 1967.
67. Early detection of health impairment in occupational exposure to health hazards, WHO Technical Report Series No. 571, World Health Organization, Geneva, 1975.
68. **Hackathorn, D., Brinkman, W., Hathaway, T., Talbott, T., and Thompson, L.,** Validation of a whole blood method for cholinesterase monitoring, *Am. Ind. Hyg. Assoc. J.,* 44, 7, 547, 1983.
69. **Ellman, G., Courtney, G., Andres, V., and Featherstone, R.,** A new and rapid colorimetric determination of acethylcholinesterase activity, *Biochem. Pharm.,* 7, 88, 1961.
70. **Winteringham, F. and Disney, W.,** A radiometric method for estimating blood cholinesterase in the field, *OMS Bull.,* 30, 119, 1964.
71. **Izmirova, N.,** Methods for determination of exposure of agricultural workers to organophosphorous pesticides, in *Field Worker Exposure During Pesticide Application,* Stud. in Env. Sci. 7, Elsevier Science Publ. S., 169, 1980.
72. **Wilhelm, K., Vandekar, M., and Reiner, E.,** Comparison of methods of measuring cholinesterase inhibition by carbamates, WHO Bull. No. 48, World Health Organization, Geneva, 41, 1973.
73. **Wilhelm, K. and Reiner, E.,** Effect of sample storage on human blood cholinesterase activity after inhibition by carbamates, WHO Bull. No. 48, World Health Organization, Geneva, 235, 1973.
74. NIOCH Pocket Guide to Chemical Hazards, USD of Health and Human Services, National Institute for Occupational Safety and Health (U.S.A.), September 1985.
75. IRPTC Data Profile on Carbaryl.

Chapter 4

ORGANOCHLORINE COMPOUNDS

I. INTRODUCTION

The organochlorine pesticides (OCPs) cover a wide range of chemical structures: (1) compounds of the cyclodiene series such as aldrin, dieldrin, endrin, heptachlor, isodrin, endosulfane, and chlordane, (2) halogenated aromatic compounds such as DDT, kelthane, methoxychlor, chlorbenzylate, and chlorphenesin, (3) cycloparaffins such as hexachlorocyclo-hexane or benzene hexachloride (BHC) and lindane, and (4) chlorinated terpenes such as polychlorcamphenes and polychloropinenes.

The best known pesticide from this group is DDT. In 1974 the world production of DDT was 60,000 tons (France, India, and U.S.). The ban or restriction of DDT in many countries is based on ecological considerations; however, in some countries it is still used in agriculture and vector control.

In treated fields DDT can be detected more than 6 months after application. DDT can be carried thousands of kilometres from treated areas and even be found in such remote areas as the Antarctic. Today DDT is still ubiquitous in the environment, due to its past wide use and its chemical and physical characteristics (Table 1). DDT persists for more than 10 years in the soil and accumulates in living organisms through the food chain (bioaccumulation or biomagnification).

Other very persistent OCPs are compounds of the cyclodiene series, particularly, aldrin, dieldrin, and endrin.

Its persistence in the environment and the adaptation of pests are the main reasons for decreased OCP use in the last two decades.

II. PROPERTIES

OCPs represent a group of pesticides with a wide variety of chemical and physical properties for individual compounds (Table 2 and 3). Their common characteristics are that almost all are solid compounds (insoluble in water but well soluble in organic solvents) and they are stable to air, light, heat, and carbon dioxide. OCPs are not attacked by strong acids, but in the presence of alkali they become unsteady and readily dechlorinated. Pesticides from the cyclodiene group are more stable toward alkalies and are susceptible to epoxidation. Oxidizing agents and strong acids attack the unchlorinated ring.

III. USES

The organochlorine compounds are used on a large scale in agriculture, forestry, and public health. In agriculture they act as insecticides, acaricides, and fumigants to control pests in orchards, vegetable, grain, cotton, and tobacco fields and vineyards. Some of them are also used for seed dressing and as rodenticides. In the field of public health they have played a decisive role in eradicating certain parasitic diseases such as malaria. They are still used in many developing countries to control transmissible diseases.

IV. METABOLISM

OCPs enter the organism by inhalation, ingestion, and dermal absorption. The hazards from dermal absorption are small when the compound is used as a dust. On the other hand, when

TABLE 1
Environmental Levels of DDT[1]

Site	Concentration in ppm
Air	0.000001
Water	0.00001
Plankton	0.0003
Water plants	0.01
Fish	0.5–10
Soil	0–2
Birds	2
Milk	0–5
Meat	0.5–2

dissolved in oil or organic solvents, it is readily absorbed through the skin and constitutes a considerable hazard.

The metabolism in the cells involves various mechanisms, such as oxidation and hydrolysis. They have a strong tendency to penetrate cell membranes and store themselves in the body fat. Due to this lipotrophic tendency, OCPs are fixed in lipid-rich cells, i.e., the central nervous system, liver, kidneys, and myocardium. In these organs, they damage the functioning of important enzyme systems and disrupt the biochemical activity of the cells.[2] OCP elimination takes place through the kidneys.

Feldmann and Maibach studied the percutaneous penetration of aldrin, dieldrin, and lindane.[3] Radio-labeled ^{14}C substances were applied to the forearms of human subjects, and their urinary excretion of ^{14}C was quantified. The urinary recovery of ^{14}C, 5 d after single i.v. administration, was studied to correct the skin penetration data for incomplete urinary excretion.

Feldmann and Maibach found that 7.8% aldrin was excreted for 5 d, as well as 7.7% dieldrin and 9.3% lindane.

Dieldrin blood levels of 0.0006 μg/ml correspond to a daily exposure of 0.1 μg/kg/d. The half-life of dieldrin in the blood of exposed workers has been estimated at 0.73 years.[4]

Toxicokynetics of pentachlorphenol (PCP) in man was studied. PCP was given orally to three volunteers at single doses of 3.9, 4.5, 9 and 18.8 mg.[5] Elimination half-life of 20 d was determined. This fact is connected with the low urinary clearance which is due to the high plasma protein binding(about 96%) and the tubular reabsorption. The increase in daily excretion was estimated by alkalization with oral intake of sodium bicarbonate.

The daily elimination of pentachlorophenol in the urine of individuals without specific exposure ranged from 10 to 48 μg/d, plasma levels being 19 to 36 μg/l.[5]

DDT in the organism is subjected to biotransformation (Figure 1). DDE is the mean metabolite found in blood and adipose tissues. After several stages of biotransformation the DDA metabolite of DDT is excreted by urine.

V. TOXICITY: MECHANISM OF ACTION

Data on the acute toxicity of OCPs are presented in Table 4.

A. MECHANISM OF ACTION

OCPs may cause liver damage. An increase in alkaline phosphatase and aldolase activity has been reported. Protein and lipid synthesizing, as well as the detoxifying and excreting functions of the liver, are affected.

OCPs induce microsomal enzyme production. The smooth endoplasmic reticulum increases so much, that the entire liver cell enlarges and hypertrophy occurs.[7-9]

TABLE 2
Chemical Structure and Names of Some Organochlorines

Structural formula	Chemical and common name
	DDT, chlorophenothane, dicophane, dichlorodiphenyl, trichloroethane
	Methoxychlor, dimethoxy-DT Dianisyl-trichloroethane
	Aldrin
	Chlordane, Octachlor, velsicol 1068
	Heptachlor
	Lindane
	Hexachlorobenzene

TABLE 3
Physical Properties

Property	DDT	Methoxychlor	Aldrin	Chlordane	Heptachlor	Lindane	Hexachlorbenzene
Melting point		89°C technical – 77°C	104°C technical – 49°C	103°C technical – 104°C	95°C technical – 46°C	113°C	229°C
Solubility in water	Insoluble	Insoluble	Practically insoluble	Insoluble	Practically insoluble	At 20°C slightly 10 ppm	Insoluble
Solubility in organic solvents	Readily in most aromatic & chlorinated solvents	Aromatic solvents	Acetone benzene xylene	Organic solvents	Benezene acetone	Acetone aromatic & chlorinated solvents	Cold alcohol benzene chloroform ether
Vapour pressure (volatility)	1.9×10^{-7} mm Hg at 20°C	—	2.31×10^{-5} mm Hg at 20°C	1×10^{-5} torr at 25°C	3×10^{-4} mm Hg at 25°C	9.4×10^{-6} mm Hg at 20°C	1.089×10^{-5} mm Hg at 20°C

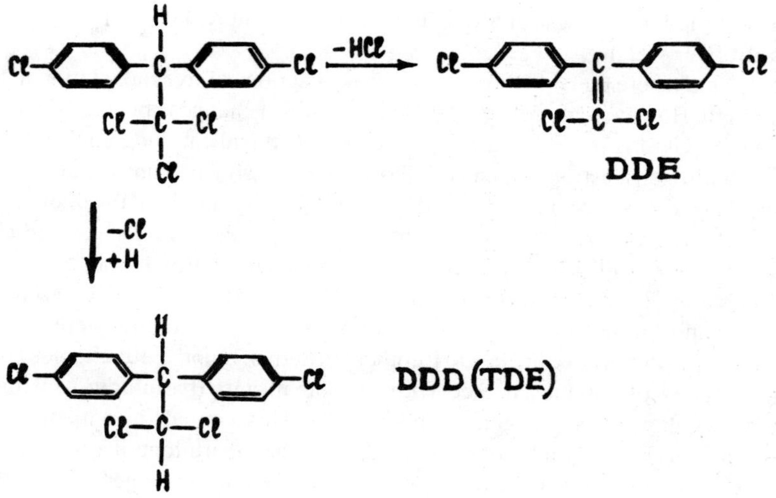

FIGURE 1. Metabolism of DDT (adapted by EHC 9).[2]

TABLE 4
LD$_{50}$ in Rats For Selected Organochlorines[6]

Compound	LD$_{50}$ mg/kg	
	Oral	**Dermal**
Aldrin	38–67	98–200
Endrin	7–18	15–18
Chlordane	335–430	690–840
Thiodane	30	359
Heptachlor	50–150	120–320
DDT	250	LC$_{50}$–150 mg/m^3 250–500 in oil 3000 as powder
Kelthane	809	
Methoxychlor	600–809	
Chlorobenzylate	960	
Lindane	88–225	900–1000
Chlorinated terpens	20	LC$_{100}$ = 70 mg/m^3 LC$_{min}$ = 20 mg/m^3
Toxaphene	45–400	500

1. Interaction Between Organochlorine Compounds, Drugs, and Hormones

There is a great number of investigations on the activity of organochlorine pesticides and their interaction with the natural products of normal metabolism in the body. An increase of the oxidative metabolism of hexobarbital, aminopyrin, and chlorpromazine has been proved.[10,11]

An apparent improvement of the tolerance toward phenylbutazone, cumarin anticoagulants, and barbiturates was induced by oral or i.p. application of 50 mg/kg chlordane or DDT for 4 d; the experimented substance was given on the fifth day. A single injection with dieldrin (40 ng/kg) or DDT (200 mg/kg) provoked a progressive increase of the degradation of these compounds (up to 200-300% of the normal rate).

This biological reaction is very sensitive. With 5 mg/kg chlordane applied orally 3 times a week, a clearly manifested effect upon the detoxication of antipyrin and phenylbutazone was obtained. Oral application of such a low dose as 1 mg/kg DDT provoked a reaction of the same type. The phenobarbital sleep time was decreased. After one injection with 1 mg/kg DDT the effect continued 2 months and disappeared during the third month.

The effect of moderate doses (of the order of 1 mg/kg dieldrin) upon liver diminishes with the increase of liver weight.

It should be noted here that the frequency of these reactions of liver adaptation in toxicology is extremely high. However, at greater OCP doses liver cell changes appear. This puts forward the question of a possible hazard of direct or indirect chemical carcinogenicity.

Studies on androgens, estrogens, and corticosteroids[7] established an acceleration of steroid metabolism. This effect accounts for a decreased fertility in animals treated with OCP formulations. During 3 weeks, 20 mg/kg chlordane was applied i.p. once a week; mouse fertility decreased. As a result of MFO stimulation, microsomal hydroxilation of estrogens and androgens increases. This metabolism stimulation is accompanied by a decreased duration of steroid-induced anesthesia and a lowered uterotropic effect of the administered estrogens.[9]

Data on the effect drugs have on the OCP content in human blood are available. The p,p'DDE content in the blood of patients, treated with anticonvulsants (barbiturate) for more than 3 months, has been lower than in general population. This content in patients treated with phenythion was 1 μg/kg, and in patients treated with phenobarbitone it was 3 μg/kg; with a combination of both drugs it was below 1 μg/kg. From the control subjects, only 1% had less than 2 μg/kg content of the same substance, while in the treated patients, the percentage has been as high as 82%.[12,13]

The authors discuss a "depesticidation" of man via drugs. Experiments with volunteers have been performed in this direction.[14]

A case is reported in support of these findings.[15] A patient, a farmer treated for 20 years with barbiturates, had less than 1 ppb DDT total, dieldrin, and heptachlor in the serum. A control population of 391 farmers with the same exposure had 11.9 ppb. The author considers two possible hypotheses to explain these data. The first is that certain microsomal enzymes were induced by the barbiturates involved in the pesticide breakdown. The second is that serum proteins bound the pesticides and they became relatively inert. Drugs compete with pesticides for the same protein binding sites. Displaced pesticides from the more tightly bound drugs will be metabolized and excreted easily. Davies et al. supported the first hypothesis, and Schoor — the second one.[12,13,15]

Later, another case of a worker treated with anticonvulsants (diphenylhydantoin and phenobarbital) in a DDT formulation plant was published.[16] A 3-year DDT monitoring in serum demonstrated very low levels (usually zero or traces), and only once were 1 ppb pp'DDT and 1 ppb pp'DDE found. Others had 20 to 30 ppb pp'DDT and 44 to 89 pp'DDE in serum. The mean serum levels for the general population were 4.4 and 14.9 ppb, respectively. This finding demonstrated an exclusively high potential of anticonvulsants for "depesticidation", even in persons exposed to much higher levels of DDT than the general population.

Similar results were obtained in mentally retarded patients treated with different drugs.[17] In another study by the same authors, baseline serum levels for pp'DDT were 4.7 ppb and for pp'DDE they were 22 ppb. In this study the levels of the control group were 5.1 ppb and 31.2 ppb, respectively. Patients receiving chemotherapy had much lower mean serum levels — 7.4 ppb pp'DDE and 1.1 ppb pp'DDT, as well as a much lower frequency of p,p'DDT occurrence. The group taking both phenobarbital and diphenyl hydantoin in combination had the lowest means for serum levels. The mean for pp'DDE was only 1.6 ppb and the range was 0.1 to 2.9 ppb. In the same group, pp'DDT was detected in only 40% of the samples, with a mean level of 0.1 ppb (0 to 5 ppb).

In experiments on dogs op'DDD caused selective adrenocortical atrophy. op'DDD was used to treat Cushing's syndrome due to bilateral cortical hyperplasia.[18] The duration of the course (3 g, then 2 g daily) was 6 months. Remission was obtained at the cost of adrenal insufficiency, affecting cortisol as well as aldosterone secretion.

B. REPORTED PROTEIN-OCP BINDING

Labeled OCP has been incorporated into a nuclear fraction of cultured human cells; human embryonic diploid cells were used.[19] Labeled aldrin, DDT, dieldrin, and HCB were dissolved in ethanol and added to the medium; the final concentration of the compounds in the medium was $4 \times 10^{-6} M$. DDT, aldrin, and dieldrin were incorporated markedly into the nuclear fraction. The rate of HCB was very low. According to author suggestions, the labeled compounds incorporated into the nuclear fraction may be due to the binding to the nuclear proteins but not the cellular DNA.

The effect of organochlorine insecticides upon immune processes has been studied.[20] DDT inhibited the synthesizing activity of the antibody-producing cells, modifying the immune reaction and decreasing α-globulins in serum proteins. The authors have found decreased participation of the adrenergic system in stress reaction. At the same time, they established an effect of the immune phenomena upon DDT cumulation compared to control animals. A decreased accumulation of hexachlorane and an increased accumulation of dieldrin was also established. The authors concluded that the accumulation of organochlorines was modified when the activity of the enzymes involved organochlorine detoxification produced an immune reaction against a foreign protein.

The process of organochlorine accumulation is influenced by thyroidectomia and adrenectomia, as well.

Experimental animal studies indicate that DDT, aldrin, dieldrin, chlordane and heptachlor may be carcinogenic to humans.[21-25]

VI. EXPOSURE: DOSE-EFFECT RELATIONSHIP

Of the greatest importance is exposing the general population and worker groups to DDT. According to Campbell 86% of the daily human intake of organochlorine insecticides occurs via food.[26] The anual intake dose is calculated up to 50 mg, from which 44.8 are absorbed with the food, 0.03 with the air, 0.01 with the water, and 5 mg via other media.

In hot countries, the absorption via food can be only up to 50% of the daily dose. The greatest significance is given to domestic dust the DDT concentrations in blood being highly dependent on its concentrations in the dwelling places.[12,27,28]

The reduction and ban of DDT in many countries reflects in the decrease of DDT in the environment, including the total diet. Thus, in Canada the total DDT dietary intake in 1973 was 0.007 mg/man/d, i.e. more than two times less than in 1971 (0.018 mg/man/d).[29] The highest levels, ranging from 2 to 104 mg/m^3 of DDT in the air, were found in places where the dust was prepared and packed (Medved et al., 1975, cit. after EHC 9).[2]

Among the other OCPs, lindane represents a definite interest, because it is used now in many countries for different purposes. Total diet studies in one country showed the daily dietary intake of lindane to be 0.06 μg/kg.[31] Seed dressing applicators have a potential dermal exposure of 54 to 81 mg/h to lindane. The respiratory exposure is 0.36 to 0.54 mg/h. Simple precautions, such as wearing gloves and respirators (when the operator is near the treatment operation) would minimize the exposure.[30]

The average daily intake of HCB in 1976 — 78, estimated from a total dust study, was 1 μg/ man (maximum value 12 μg/man) in the Netherlands.[32]

In 1980 HCB residues in fat samples from meat and poultry in U.S. were found in 3 to 6% of the samples. Of these, 95% contained less than 0.1 ppm, and only 7 out of more than 25,000 samples contained HCB levels higher than 0.5 ppm.[33]

In Czechoslovakia HCB was found in all food chain links.[34]

For more than 10 years, the results of HCB monitoring in West Germany showed a steady decrease of the HCB concentration in cow milk, from >0.11 to about 0.02 mg/kg milk fat.[35] Studies on human milk indicated a slight tendency toward increased HCB, from about 0.3 to more than 0.6 mg/kg/fat.

In Israel HCB residues in milk were found in almost all samples on a fat basis of 0.08 ppm (0.01 to 0.7 ppm).[36]

Total diet studies in two countries demonstrated an average daily dieldrin intake of 7 µg/man (2.7 to 22 µg/man).[4]

Chlordane residues seldom occur, and the uptake by man is negligible.[37]

During heptachlor spraying air concentrations of 0.6 to 1 mg/m³ have been measured.[38]

Chlordane was found in the household dust of farm homes (mean level 5.79 mg/kg air dried dust).[37,40] The chlordane residues in meat were 0 to 106 µg/kg, milk — 0.02 to 0.06 mg/l, and eggs — 2 µg/kg.[37]

A. OCP LEVELS IN HUMAN TISSUES

The content of organochlorine substances in tissues, their biological effect, and the influence the physiological and pathological states of the organism have upon their accumulation are studied. By means of gas chromatography with electron capture detection, all organochlorine compounds and their metabolites now in use can be practically identified. Summarized data for OCP content in human tissues are available.[2,39]

1. Organochlorine Pesticide Content in Fat Tissue

The organochlorine pesticide content is best studied in fat tissue, because in it they accumulate in the greatest quantities. The variations for the different countries are a function of the local application conditions. Considerable variations between the individuals of the same country also exist. According to data from 22 countries given in Table 5, the average values for the different organochlorine pesticides are DDT-total from 1.75 to 30, HCH-total from 0.16 to 2.43, dieldrin from 0.046 to 0.68, and heptachlorepoxide from 0.0085 to 0.19 mg/kg. In countries with a hot climate, the levels are higher. DDT was found in the fat tissue of Eskimos at a concentration of 3.1 mg/kg.[40]

Sex differences in the pesticide content are generally insignificant.[41] However, some authors found differences in DDT accumulation in the different races, as well as higher quantities in men than women.[42,43]

The results of an investigation in the U.S. during the years 1967 and 1968 show higher concentrations of DDT residue in the black population.[27]

According to some authors, OCP accumulation increases with the age. The greatest quantities have been determined in persons over 45.[43,45] In Iceland, higher quantities of DDT and its metabolites were found in children.[28]

The OCP content also depends on a complex of factors related to their application. Thus, the levels of lindane in fat were about 0.42 ppm in a European country and 1.43 ppm in an Asian country.[31]

The interaction between the different factors and the accumulation of pesticides in human tissue is studied in detail.[87] It is interesting to trace out the evolution of human pesticide burden in the course of several years. According to the common opinion, there is no increase in the OCP content of human tissue due to the limitation of its use. Some authors support the fact that in U.S. there have been no differences in OCP (DDT) quantity in the fat tissue of the general population since 1951.[88] This conclusion confirms the opinion that a balance between absorbed and eliminated doses is established under condition that only insignificant changes occur in the environment and the accumulation of DDT and its derivatives in the organism persists. Other authors have come to the same conclusion concerning the population in Arizona.[87] However the data are considered insufficient to confirm this conclusion.[27]

The cumulative effect of OCPs has been studied on volunteers, the doses exceeded the

TABLE 5
Concentrations (mg/kg) of Organochlorine Pesticides in Adipose Tissue of General Population — Various Countries

Pesticide	Country	Concentration		Author
Total DDT Dieldrin	Australia (Melbourne)	1.8 0.046		Bick (1967)[45]
Total DDT BHC Dieldrin	Australia (West)	9.3 0.68 0.68		Wassermann (1968)[46]
Total DDT Lindane Dieldrin	United Kingdom	3.9 0.015 0.21		Robinson et al. (1965)[47]
Total DDT BHC Dieldrin	United Kingdom	2.84 0.19 0.34		Cassidy (1967)[48]
Total DDT BHC Dieldrin	United Kingdom	3 0.31 0.21		Abbott et al. (1968)[49]
Dieldrin	United Kingdom	0.27		Hunter et al. (1967)[50]
Total DDT BHC Dieldrin	Argentina	13.17 0.4 0.19		Wassermann et al. (1969)[43]
Total DDT BHC Dieldrin Heptachlorepoxide	Argentina	4.63 1.53 0.07 0.24		Fernandes et al. (1971)[51]
DDT	Bulgaria	10.6 14.7		Kaloyanova et al. (1972)[52] Rizov (1977)[53]
Total DDT Total DDT	South Africa FRG	6.38 2.2		Wassermann et al. (1970)[59] Maier-Bode (1960)[55]
DDT DDE	GDR	3.7 8.5		Engst et al. (1967)[56,57]
Total DDT DDE	Israel	19.2 8.5		Wassermann et al. (1965)[44]
Total DDT	persons aged: from 0 to 9 yrs from 10 to 99 yrs	10.2 18.1		

Pesticide	Country	men	women	Author
Total DD Dieldrin PCB	Israel	9.94 0.38 2.45	4.62 0.21 2.45	Wassermann et al. (1973)[58]
Total DDT DDT DDE Dieldrin	India (Delhi)	31 20.4 10.7 0.03		Dale et al. (1965)[42]
Total DDT DDT DDE Dieldrin	India (soldiers)	12 7 4 0.06		
Total DDT	India	13.8 (0.17–176.5) in 43% up to 25		Ramachandran et al. (1974)[59]

TABLE 5 (continued)
Concentrations (mg/kg) of Organochlorine Pesticides in Adipose Tissue of General Population — Various Countries

Pesticide	Country	Concentration		Author
Total DDT	Spain	14.8		Martines, Wassermann (1967)[60]
DDT		6.5		
DDE		8.3		
DDE	Italy	7.43		Paccagnella (1967)[61]
Lindane		0.01		
Total DDT		15.48		Del Vecchio, Leoni (1967)[62]
Dieldrin		0.88		
Total DDT	Canada	4.9		Read et al. (1961)[63]
Total DDT	Canada	9.22		Mastromateo (1971a)[64]
		men	**women**	
Total DDT	Poland	11.08	9.83	Pomorska, Szuchki (1971)[65]
DDE	Poland	7.28		Juskeewiez, Stec (1971)[66]
DDT		4.08		
BHC		0.13		
Total DDT	U.S.S.R.	3.3		Vaskovskaja, Komarova (1967)[67]
		0.200–15		Grasheva (1968)[68]
Total DDT	U.S.	6.69		Dale, Quiby (1963)[69]
BHC		0.2		
Dieldrin		0.15		
Total DDT	U.S.	10.3 (0.900–64.1)		Hoffmann et al. (1964)[70]
Lindane		0.56		
Dieldrin		0.11		
Total DDT	U.S.	10.56		Fisherova-Bergerova (1967)[41]
DDT		2.81		
DDE		6.67		
TDE		0.28		
Dieldrin		0.22		
Total DDT	U.S.	9.6		Hoffmann et al. (1967)[71]
BHC		0.48		
Dieldrin		0.14		
		white	**black**	
Total DDT	U.S. (Florida)			Davies et al. (1968)[12]
	Age: 0–5 yrs	5.5	7.9	
	Over 5 yrs	8.4	16.7	
		white	**black**	
Total DDT	U.S.			Mrak (1969)[27]
	Age: 0–5 yrs	4.5	6.1	Draham (1965)[72]
	41–50 yrs	6.08	12.14	
	81–90 yrs	6.2	14.89	
Dieldrin	0–5 yrs	0.11	0.12	
	41–50 yrs	0.12	0.11	
	81–90 yrs	0.1	0.15	
DDT	U.S.	1.43		Morgan, Roan (1970)[73]
DDE		4.63		
Dieldrin		0.13		
Mirex	U.S.	0.16–5.94		Kutz et al. (1973)[74]
DDT	U.S.	8		Diechman (1973)[75]
Dieldrin		0.18		
Endrin		0.02		
Lindane		0.42		
Heptachlor		0.17		

TABLE 5 (continued)
Concentrations (mg/kg) of Organochlorine Pesticides in Adipose Tissue of General Population — Various Countries

Pesticide	Country	Concentration		Author
Total DDT	Tailand			Wassermann et al. (1972)
	age: 5–24 yrs	8.2		
	25–44 yrs	12.4		
	over 45	6		
Dieldrin,				
Heptachlor, BHC		below 0.2		
Total DDT	Uganda			Wassermann et al. (1974)[74]
	age: 0–5 yrs	2.3		
	5–24 yrs	3.84		
	25–44 yrs	2.9		
	over 45 yrs	2.4		
BHC, Dieldrin		below 0.1		
Heptachlorepoxide		below 0.2		
Total DDT	Hungary	12.4		Denes (1962)[78]
Total DDT	France	5.2		Hayes, Dale (1963)[79]
DDT		1.7		
DDE		3.5		
DDE	France	5.2		Fournier (1970)[80]
DDT		1.9		
BHC		1.15		
Heptachlorepoxide		0.15		
Total DDT	Netherlands	1.75		De Vlieger et al. (1968)[81]
Dieldrin		1.17		
Lindane		0.1		
Heptachlorepoxide		0.0085		
Total DDT	Czechoslovakia	9.2		Halacka et al. (1965)[82]
DDE		3.7		
DDT		5.5		
Total DDT	Czechoslovakia	20.34		Rosival et al. (1970)[83]
BHC		9.78		
Total DDT	Yugoslavia	men	12.3	Adamović et al. (1970)[84]
		women	10.3	
Total DDT	Japan	8.2		Kasai (1970)[85]
DDT		0.17–5.5		
DDE		0.6–3.7		
Total DDT	Japan	4.2		Tatsumi (1972)[86]

quantities ingested with food. The effect of doses from 0 to 35 mg/man/d DDT applied during 2.5 to 5 months, pure or in milk emulsion, was studied.[90] The DDT that accumulated in the fat tissue was proportional to the dose applied, and the distribution of DDT and DDE varied with the time (65% DDT in the beginning of the experiment, 14% at the end of the 18-month period). In the persons with an absorption rate of 0.5 mg/kg/d, the accumulation in fat tissue has been from 65 to 129 mg/kg. The fat tissue of these subjects received the highest dose, containing 105 to 619 ppm. The average dosages of p,p'-DDT and of the p,p'-DDT related compounds were respectively 1250 and 555 times greater than those in general population.

The derivatives of DDT degradation in the liver (20% in the form of DDA) are excreted in the urine.

The relation between the average DDT concentration in fat tissue and the daily quantity of the DDT intake has been calculated as follows:[91]

$$Log_{10}C = 0.0124 + 0.6539 Log_{10}(10^2 \times \text{daily DDT dose in mg}); (\pm 0.105)$$

where C is the DDT concentration in fat tissue and ±0.105 is the standard error.

The existence of a balance between the dose received and the content in the fat tissue was evidenced by an experiment using dieldrin on volunteers.[92,93] In the course of 24 months, three groups of subjects received 10, 50, and 211 µg/d doses of dieldrin. A dependence was observed between the received dose and the concentration of the chemical in blood and fat tissue.

Based on the above mentioned works, a calculation was performed on the daily dose per man intake in different countries. Canada had up to 8.6 µg/d, England had up to 10.5 µg/d, the U.S. had up to 19.9 µg/d, and India had up to 1.6 µg/d.

The doses received by workers engaged in dieldrin production are found to range from 570 to 1186 µg/man/d.

2. OCP Content in Different Tissues, Blood, and Urine

As it was already underlined above, the absorption and accumulation of OCP on one side, and their metabolism and excretion on the other side, develop in such a way, that a balance between their content in the tissue and the environment is established. OCPs are not in a constant quantity in the fat tissue since a dynamic balance with the other organs is developing, mediated by the blood.

The OCP concentrations in the kidneys, brain, and endocrine glands are about 100 times lower than in fat tissue, and they are about 10 times lower than in the liver (Table 6). The proportion of DDT in blood, fat tissue, and the liver is 1:306:27.

Comparative studies on OCP content in serum and fat tissue were performed on 200 subjects.[99] The authors did not find a direct relationship between the DDT content in serum and fat tissue. They are of the opinion that, based on the DDT levels in serum, it is not possible to predict its content in lipids.

The values found in the blood of the general population are shown in Table 7.

In children the quantities of p,p'DDT are higher than of p,p'DDE.[51] The authors interpret this as an indication of a lower capacity for metabolizing DDT in children.

In one study a DDT metabolite known as DDA has been found in quantities less than 0.02 to 0.35 mg/dm³ in 26% of the urine samples.[91] From the other DDT metabolites, a frequent finding is DDE (Table 7.) The total DDT excretion in urine for the general population is 0.02 mg/man/d.[100] OCPs are excreted in bile as well.

B. OCP AND PREGNANCY

OCPs pass through the barrier of the placenta in the tissues of the embryo (Table 8), where they can be metabolized.[101]

The quantities measured in a great number of embryos are within the limits of those established for the general population, but they are lower than the maternal ones.[102,103] It could be supposed that they will have an effect upon the embryo in cases of more significant maternal exposure.

The concentration of DDT and its metabolites was determined in the placenta and venous blood of 9 women; their pregnancies resulted in stillbirths, and 2 women had dystrophic changes in the embryo.[104] In the early period of the pregnancy all investigated women had been exposed to DDT. The residual quantities in the venous blood and placenta have been significantly higher than in the controls. Mother's milk contains 13 times more total DDT, 6 times more BHC, and 8 times more dieldrin, than blood.

Sucklings are exposed to the OCPs contained in mother's milk (Table 9). According to Dofrtoth (cited after Mrak) they intake 0.02 mg/kg/b.w. DDT.[27] This quantity represents a

TABLE 6
Concentrations (in kg/mg) of Organochlorine Pesticides in Various Tissues of the General Population

Tissue	DDE	DDD	Total DDT	DDT	Dieldrin	BHC	Heptachlor epoxide	Author
Liver	.35	0.34	0.12	0.889	0.035	—	—	Fisherova-Bergerova et al. (1967)[41]
Kidneys	0.077	0.017	0.036	0.141	0.013	—	—	
Brain	0.123	—	traces	—	0.035	—	—	
Gonads	0.059	—	traces	—	0.035	—	—	
Brain	0.12	—	0.04	—	0.04	—	—	
Liver								Radomski et al. (1968)[97]
Normal	0.35	0.36	0.10	—	0.03	—	—	
Portal cirosis	1.19	0.2	0.53	—	0.10	0.12	0.05	
Malignant metastasis	2.44	0.22	0.08	—	0.07	0.1	0.03	
Toxic hepatitis	0.65	0.14	0.44	—	0.03	0.12	0.04	Casaretti et al. (1968)[98]
Bone marrow	2.08	0.076	0.411	—	0.062	—	0.004	
Kidneys	0.209	0.0022	0.0827	—	0.0056	—	0.0009	
Liver	0.200	0.0326	0.467	—	0.285	—	0.0019	
Brain	0.083	0.002	0.0105	—	0.0031	—	0.0002	
Lungs	0.0585	0.0009	0.0147	—	0.0022	—	0.0003	
Spleen	0.0305	0.0031	0.0112	—	0.0021	—	traces	
Gonads	0.0688	0.0015	0.015	—	0.0021	—	0.0001	
Mesentery	4.40	.047	1.35	—	0.063	—	0.032	
Liver	—	—	—	0.11	0.03	—	—	De Vlieger et al. (1968)[81]
Brain								
White matter	—	—	—	0.025	0.0061	—	—	
Grey matter	—	—	—	0.02	0.0047	—	—	
Liver								Morgan, Roan (1970)[73]
acute disease	0.506	—	0.114	—	0.047	—	—	
chronic disease	0.357	—	0.089	—	0.016	—	—	
Brain								
Acute disease	0.083	—	0.02	—	0.007	—	—	
Chronic disease	0.085	—	0.023	—	0.006	—	—	
Kidneys								
Acute disease	0.135	—	0.031	—	0.014	—	—	
Chronic disease	0.132	—	0.030	—	0.008	—	—	

TABLE 7
Concentrations (in mg/kg) of Organochlorine Pesticides in Blood and Urine of the General Population

DDE	DDT	Total DDT	DDA	Dieldrin	BHC	Author
Blood						
0.0194	0.0068	—	—	0.0014	0.0031	Dale et al. (1966)[105]
0.009	0.002	—	—	—	—	Nachman (1969)[106]
0.009–0.027	0.0025–0.0055	—	—	—	—	Apple et al. (1970)[107]
0.00436–0.0548	0.0026–0.0639	—	—	—	—	Mrak (1969)[27]
0.033–0.028	0.007–0.011	—	—	0.004	—	Morgan, Roan (1970)[73]
—	—	0.035–0.110	—	—	—	Long (1971)108
0.0157	0.004	—	—	0.0015	0.0014 U.S.	Radomski et al. (1971)[109]
0.0145	0.0032	—	—	0.0014	0.023 Arg	
—	—	0.0167	—	—	0.0168	Kasai (1972)[110]
Urine in 1						
0.0156	0.007	0.0244	0.0142	—	—	Cueto et al. (1967)[100]
0.00042	0.00036	—	0.01–0.18	—	0.0142	Granmer et al. (1969)[110]
—	—	—	—	—	0.01–0.18	Durham et al. (1965)[91]

TABLE 8
Concentrations (in mg/kg) of Organochlorine Pesticides in Fetal Tissues and New-Born Babies

Tissue	Total DDT	Total BHC	Dieldrin	Heptachlor epoxide	Author
Embryonic tissue	1.607	—	0.045	0.06	Zavon (1967)[111]
Adipose tissue	5.68	—	0.17	—	Fisherova-Bergerova (1967)[41]
Liver	0.155	—	0.007	—	
Kidneys	0.085	—	0.005	—	
Brain	0.005	—	0.005	—	
Gonads	0.005	—	—	—	
Embryonic tissue	0.59–3.03	—	—	—	Curley et al. (1968)[112]
Blood from new-born babies	0.008	0.0012	0.0013	—	Polishuk et al. (1970)[101]
Blood from new-born babies	0.0073	0.053	0.00059	0.00006	Radomski et al. (1971)[109]
Embryonic tissue	1.04 subcutaneous adipose tissue 0.58 liver				Komarova et Vaskovskaya (1968)[113]
Placenta	Traces up to 5 in 45/150 samples				
Adipose tissue	3.1				
Blood	Traces — 1.0 in 28/40 samples				Komarova (1973)[114]

TABLE 9
Concentrations (in mg/kg) of Organochlorine Pesticides in Human Milk

Total DDT	Total BHC	Dieldrin	Heptachlor epoxide	Author
0.075–0.17	0.008–0.015	0.001–0.019	—	Egan et al. (1965)[115]
0.12	—		—	Heyndrickx, Mayes (1969)[116]
0.05–0.37	0.018 (α-isomer)	—	—	Mrak (1969)[27]
0.0062–0.078	0.0024–0.0098	0.0015–0.007	0.001–0.0027	Curley, Copeland, Kimbrough (1968)[112]
Lipid fraction				Mastromateo (1971a)[64]
2.6	—	0.13	—	
0.064	—	—	0.046	Kasai (1972)[110]
0.036	—	—	—	Tatsumi et al. (1972)[117]
0.25	0.0003	—	—	Juszkiewicz et al. (1972)[117]
0.230–1969	—	—	—	Engst u. Knoll (1972)[56]
0.160–1970				
0–0.27				Hornabrook et al. (1972)[118]
0.112	0.018 (β-isomer)	—	—	Acker, Schulte (1970)[119]

fourfold excess of the allowable daily dose recommended by the WHO. Heptachlor was less than 2.5 mg/kg and oxychlordane and chlordane had mean values of 0.096 mg/kg in milk fat.[120,121]

Large scale studies to assess human exposure to selected organochlorine pesticides through biological monitoring have been organized by the UNEP/WHO in the framework of the global Environmental Monitoring System (GEMS). The project was carried out during 1981 and 1982. Eleven countries from four continents participated in the project (Belgium, Egypt, West Germany, India, Israel, Japan, Mexico, the People's Republic of China, Turkey, the U.S., and Yugoslavia).[122]

Selected OCPs were determined — mainly DDT and metabolites and β-HCH, — as well as PCBs. The reported values are shown in Table 10. The relatively high values in India, China, and Mexico are certainly due to the continued use of DDT in vector control and agriculture. In contrast to that, the levels of PCBs in human milk fat were higher in European countries and Japan.

The intake of organochlorine compounds by the breast-fed infant were calculated based on the assumption that the infant consumes 130 g milk/kg/b.w. daily. The calculated daily intake

TABLE 10
Organochlorine Compounds in Breast Milk (means) in mg/kg/fat

Country	p,p′DDT	p,p′ DDE	HCH	PCBs
Belgium	0.13	0.94	0.2	0.81 (0.10–2.3)
China	1.8	4.4	6.6	Not detected
FRG	0.25	1.2	0.28	2.1 (0.24–10)
India	1.1	4.8	4.6	Not detected
Israel	0.23	2.2	0.29	0.45 (<0.5–2.1)
Japan	0.21	1.5	1.9	0.35–0.51 (0.10–0.98)
Mexico	0.71	3.7	0.4	Not detected
Sweden	0.09	0.81	0.085	0.97–1.3 (0.40–2.4)
U.S.	<0.1 (<0.1–3.1)	1.6	<0.05	<1 (<1–5)
Yugoslavia	0.18	1.9	0.28	0.63 (0.3–1.7)

TABLE 11
Concentrations (in mg/kg) of Organochlorine Pesticides in Various Tissues of Pregnant Women

Tissue	Total DDT	Total BHC	Dieldrin	Heptachlor epoxide	Author
Placenta	0.024	0.0012	0.00148	0.0021	Lesby et al. (1969)[124]
Blood	0.0018	0.00048	0.00026	0.00024	
Plasma	0.001–0.02	0.001–0.0097	0.0001–0.006	0.0001–0.0033	Curley, Kimbrough (1969)[102]
Adipose tissue	12.6	0.035	0.085	—	Polishuk et al. (1970)[101]
Placenta	0.074	0.008	0.0065	—	
Uterus	0.029	0.010	0.0009	—	
Blood	0.018	0.0015	0.0016	0.00023	
Blood	0.014	0.013	0.0016	0.00023	Radomski et al (1971)[109].

indicates that the DDT total intake by some of mostly breast-fed infants exceeded ADI (5 µg/ kg/b.w.). In some countries the ADI is exceeded severalfold by most children.[122,123]

During pregnancy Polishuk et al. have observed quantitative changes in the accumulation of organochlorine substances and their metabolites in mother.[101] DDT and its metabolites, BHC and dieldrin, are contained in the fat tissues of pregnant women in smaller quantities than in other women (Table 11). The authors explain this phenomenon with the more active metabolism in pregnant women. Their results are consistent with the data of Curley and Kimbrough who found lower concentrations of organochlorine substances in the plasma of pregnant women; this decreased quantity was better manifested after delivery.[102] The authors are of the opinion that a possible cause of such a decrease are the physiological changes: increased progesteron and estrogen, increased water in the organism of pregnant women, hypervolemia, and drugs used during pregnancy.

C. HUMAN PATHOLOGY AND ORGANOCHLORINE COMPOUNDS

The eventual role of OCPs in the development of various diseases has been studied by a number of authors. Hofman et al. have shown that in 994 cases with different diseases from 4 Chicago hospitals, no dependence exists between the OCP content in fat tissue and the pathological changes.[71] Radomski et al. proved that a patient who died from portal cirrhosis, cancer of different organs, and arterial hypertonia, had higher concentrations of organochlorine substances, but that does not support a casual dependence between the diseases and pesticide content.[97]

Casaretti et al. have observed higher OCP levels in the tissues of patients with cachexia, cancer, and liver impairments.[98] The authors are of the opinion that the manifested decrease of subcutaneous fat in these cases was possibly responsible for the increased OCP concentrations in other tissues.

Morgan and Roan compared the OCP content in the tissues of patients who died from acute and chronic diseases. They established higher OCP content in the issues of chronic patients.[73] Wassermann et al. have come to a similar conclusion concerning fat tissues.[44]

No increase of blood dieldrin concentration was observed under surgical stress and at deliberate weight loss.[129] Robinson pointed to the risk of the insecticides moving in the blood as a result of physiological stress.[94]

D. LEVELS OF OCP IN WORKERS VS. GENERAL POPULATION AND EVALUATION OF THEIR SIGNIFICANCE

To evaluate the hazard of the accumulated OCP quantities in tissues of the general population, the comparisons are often made with the OCP content in the tissues of exposed workers (Table 12). In some cases the maximum OCP concentrations in the fat tissue of exposed workers are 10 to 100 times higher than those of the general population.

A comprehensive analysis and comparison of the data for general population and workers was performed by Laws et al.[125] In workers with high occupational exposure to DDT the authors found the mean values for p,p'-DDT in fat of men from different exposed groups to be from 39 to 128 times higher than those found in the general population of the U.S. Mean values for total DDT were only 12 to 32 times greater. For blood, the levels of total DDT were 3 to 10 times the means for the general population. The mean values in the urine were 28 to 78 times the corresponding means for the general population. DDA was the most excretory form of DDT in the occupationally exposed persons, but it was less important in the general population.

The storage range for the sum of DDT isomers and metabolites in the fat of workers was from 38 to 647 ppm, compared to the value of 8 ppm for the general population. It was estimated that the daily intake of DDT by 20 workers varied between 17.5 and 18 mg/man/d, as compared to an average of 0.04 mg/man/d in the general population. Workers store a smaller proportion of DDT-related materials in the form of DDE.

There was a significant correlation (r = +0.64) between the concentration of total DDT in the fat and serum of the workers — the concentration in fat was 338 times greater than that in serum.

Griffith and Duncan compared serum organochlorine residues in citrus workers with the national health and nutrition examination survey samples.[131] A total of 567 samples were analyzed for the presence of 10 different organochlorine insecticides. Samples were collected from field workers during the spray season. Only serum DDE values were found to be significantly higher in harvest season workers — 31.4 ppb and in spray season workers — 23.2 ppb, compared with 18 ppb in the control population.

Mean values of HCB in tissues of workers exceeded several times that of the general population.[95,96]

HCB in whole blood samples of the general population (West Germany) were 2.3 µg/l (0.05 to 13.5) in 1977 and 3.55 µg/l (0.95 to 16.3) in 1982. Wine growers exposed to HCB showed a range of 0.86 to 29.6 µg/l, with a mean of 7.35 µg/l. In autopsy material, age-dependent HCB concentrations found in adipose tissues were increasing with age (0.14 to 45.4 µg/g extracted lipids).[95] In adipose tissues of the U.S. general population, the estimated average residue level of HCB is 0.053 ppm (ranging from "not detected" to 4.33 ppm) during the period 1974 to 1983.[96]

The dieldrin concentrations in workers are from 6 to 250 times higher than in the general population. In the subcutaneous fat of volunteers with 211 µg daily intake of dieldrin, a concentration of 2.85 mg/kg was established, which is 10 times more dieldrin residue content than in the general population. At such concentrations, symptoms of intoxication were not observed.[93,94,132,133]

DDT quantity exceeding 200 times that normally contained in food is not considered hazardous, even if it is taken in the course of a period up to 18 months.[38] The DDT accumulated

TABLE 12
Organochlorine Pesticides (mg/kg) in Adipose Tissue, Blood and Urine of Workers
Exposed to Organophosphorous Pesticides

Pesticide	Conditions of exposure	Concentration in			Author
		Adipose tissue	Blood	Urine in 1	
Total DDT	Massive exposure in a plant	1131 max	—	0.02–0.35	Durham (1965)[72]
Total DDT	18 mg/d per person	38–647	—	—	Laws et al. (1967)[125]
Total DDT		18.39	—	—	Dale and Quinby (1963)[69]
BHC		0.36	—	—	
Dieldrin		0.27			
DDT	Massive exposure	—	0.43–0.129	—	Edmundson
DDE		—	0.29–0.025	—	et al. (1969)[126]
DDA		—	—	0.038–0.3	
Total DDT	Aerial spraying	—	0.025–0.041	—	Edmundson
	Plant for pesticides	—	0.017–0.064	—	et al. (1969)[127]
	Application in agriculture	—	0.032–0.041	—	
	Spraying of flowers	—	0.010–0.036	—	
	Controls	—	0.007	—	
DDT	Workers handling pesticides	—	0.025–0.133	—	Perron and Barrentine (1970)[128]
	Office personnel	—	0.008–0.046	—	
	Controls (all persons examined have been exposed to DDT at domestic use for mosquito control)	—	0.005–0.03	—	
Dieldrin	Workers exposed for 1 year	9.5	0.041	—	Hunter et al. (1967)[92]
	Workers exposed for 3–12 years	—	0.067–0.102	—	
	General population	—	0.004	—	
Aldrin		—	0.0006	—	Hunter and Robinson (1968)[129]
Dieldrin		5.67	0.0185	0.0242	
Total DDT		—	—	0.0349	Cueto and Biros (1967)[100]
DDE		—	—	0.0212	
DDT		—	—	0.0113	
Dieldrin	Low exposure	—	—	0.0023	
	Moderate exposure	—	—	0.0053	
	Considerable exposure	—	—	0.0514	
Dieldrin	Various professions	—	0.096–0.29	—	Jager (1970)[130]
	Intoxicated workers	—	0.331	—	
Total DDT	Pilots, agricultural aircraft		0.038		Yanchev and Burdarov[139]

in the tissues gradually metabolize into DDE-derivative, which is less toxic. In the course of one year, the DDT and DDE concentrations in the tissues attain a stable level, the quantity of metabolized and eliminated substances become equal to the doses absorbed.

An analysis of the results of long-term experiments upon volunteers has been performed.[90] A number of clinical indices were followed up in volunteers during a period of 4 years without any data for adverse health effect being established, although the DDT content in adipose tissue was 1000 times over that of the general population. For this reason, the authors supported the view that, in spite of DDT accumulation, these doses do not represent a hazard for the health of population.

No morbid manifestations have been related with the residues found in the general population. The only abnormal deviations found in subjects with severe occupational exposure and serum levels of 573 μg/kg DDT and 506 μg/kg p,p′DDE, were a 56% increase of the cortisol metabolism and a 19% decrease of the phenylbutazone half-life in the serum of adult subjects (Poland et al., cited after Davies and Edmundson).[13]

In other investigations on workers exposed to comparatively high DDT levels, no changes of the cortisol metabolism in the blood, nor disturbances in its secretion by suprarenals, were found.[134] Analogous studies have been performed on agricultural workers.[135] No deviations were found in the excretion of corticosteroids in urine, nor in the cortisol concentration in the serum.

The DDT concentrations in the fat tissue of rats which could stimulate the microsomal enzyme activity are of the order of 10 to 12 mg/kg. It is supposed that an increased metabolism of different drugs and steroids could be expected in the general population as a result of the accumulation of these pesticides in human tissues.

It seems that adipose tissue is not injured by the presence of organochlorine pesticides. The adipose tissue serves as a defensive system of the body for the accumulation of these compounds, which are released by the organism in small quantities and in such a way, that it can cope with their detoxication.

It is difficult to compare the reactions of persons occupationally exposed to OCPs with the ones of the general population. A great caution is necessary in performing such a comparison. Among the general population there are many unhealthy people and children with earlier exposure to pesticides, before their birth or in the period of their growing up. The sensibility toward the effect of pesticides is varying with the age.[136]

For many organochlorine preparations it is not possible today to determine the maximum allowable concentration in tissues and biological media. As exposure indices for total DDT Hunter proposed 20 mg/kg in adipose tissues and 0.10 mg/kg in blood, and for dieldrin 0.5 and 0.035, respectively.[137] This proposal is based only on the toxic action of these substances as a criterion; their other biological effects are not taken into consideration.

For monitoring exposure, a WHO study group recommended the following tentative guidelines: for total DDT in blood up to 0.2 µg/ml, no harmful effects should be expected and for 0.2 to 0.5 µg/ml action is required, such as repeated examination after an appropriate interval and appraisal of general and individual work situations. Exceeding the upper value of this level should be prevented.

For monitoring aldrin and dieldrin exposure the WHO working group recommended three levels of their residues in blood: (1) up to 0.1 µg/ml values have no harmful effect; (2) 0.1 to 0.2 µg/ml values require action and should not be exceeded; (3) more than 0.2 µg/ml values require remedial action, including temporary removal from further exposure and appraisal of work situation.[138] At the level 0.15 µg/ml there was no health impairment nor signs of enzyme induction.

VII. EFFECTS ON HUMANS

A. ACUTE INTOXICATIONS WITH OCPS

OCPs are cumulative substances, and chronic intoxications are more common than acute ones during their application and production. Only aldrin, endrin, dieldrin, and toxaphene are implicated most frequently in acute poisoning. The delayed onset of symptoms in severe acute intoxications is about 30 min. With an OCP of lower toxicity, the delay is several hours but not more than 12 h. The intoxications are marked by gastrointestinal symptoms: nausea, vomiting, diarrhea, and stomach pains; however, the basic syndrome is cerebral: headache, dizziness, ataxia, and paraesthesia. Tremors gradually set in, starting with the eyelids and the face muscles and descending toward the trunk and limbs. In severe cases, this progression leads to fits of tonoclonic convulsions, which gradually extend to different muscle groups. The convulsions may be connected with elevated body temperature, and unconsciousness and death may result. Acute intoxications may lead to bulbar paralysis of the respiratory center and the vasomotor centers, causing acute respiratory defficiency (apnea). Many patients develop signs of toxic hepatitis and nephropathy. After these symptoms disappear, some patients develop signs of prolonged toxic polyneuritis, anemia, and hemorrhagic diathesis connected with the impaired thrombocytopoiesis. Typical for toxaphene is an allergic bronchopneumonia, liver damage, and optic nerve damage.

Acute intoxication with some OCPs may take up to 72 h. When parenchymatous organs are damaged, the illness may continue for several weeks. The prognosis is usually favorable, but it may be complicated in cases of liver and kidney damage.

Eskenasy reported a case of acute poisoning; a woman ingested 27 mg/kg/b.w. DDT and 18 mg/kg/b.w. lindane.[139] Within 20 min status epilepticus developed and lasted 13 h. Intensive sweating, hepatic involvement as revealed by dyspeptic syndrome, increased transaminase activity, and modified protein balance were also demonstrated. An electroencephalogram showed slow theta- and delta-wave dysrhythmia, which was accompanied by the epileptic discharge of symmetric and bilateral synchronous appearance at 3 Hz frequency, 150 to 200 μV amplitude, and 3 to 90 s duration.

Kiuge and Olbrich reported a case of acute poisoning with combined solution of DDT and lindane in a 2-year-old child.[140] The intoxication has been manifested with abundant vomiting, dizziness, unconsciousness, convulsions in the lower limbs, and an EEG finding of subcortical distrophy.

B. CHRONIC EFFECTS OF OCP

The nervous system is among the target organs for OCPs. Polyneuritis, encephalopolyneuritis and astenovegetative syndromes have been reported after occupational exposure to OCPs.

An examination of subjects working with OCP compounds (hexachloran, polychlorpinen, hexachlorbutadien, and DDT) revealed signs of diencephalon pathology in 2.8% of the cases, which was often accompanied by other symptoms of chronic intoxication.[141] Most often a vegetative form of diencephalon pathology was observed, with manifestations of headache, giddiness, paraesthesia in the extremities, vascular lability, and neurocirculatory disturbances (hypertension, acrospasm, and cyanosis). Less frequent was the vegetative-visceral form, manifested by coliclike pains in the umbilical and rib region and by dyskynesia of the bile ducts.

Both the vegetative-vascular and vegetative-visceral forms of diencephalon pathology were observed, not only in overt chronic OCP intoxications and long-term exposure, but also in subjects with a rather moderate employment duration — up to 3 years, especially in disinfectors exposed to a complex of pesticides.

The diencephalon disturbances are distinguished by some peculiarities when they have been provoked by a continuous contact with pesticides. On the background of the characteristic signs of chronic intoxication, disturbances of metabolic and endocrine character appear and develop, reminding in such cases the Itzenko-Cushing syndrome.

All OCPs are central nervous system stimulants, producing convulsions that are frequently epileptic; abnormal EEG data have been reported. In investigations of workers continuously exposed to OCPs (a total of 73 subjects), 21.9% gave abnormal EEG data, the changes were most expressed at a length of labor activity from 1 to 2 years. In the groups of shorter or longer exposure less significant changes were observed.

Different EEG changes were found. In some cases, bitemporal spike waves with changeable lateralization, low voltage, and diffuse theta activity were observed. In other cases, a paroxismal radiation consisting of slowspike waves, spike wave complexes, and rhythmic peaks of low voltage were registered. Deviations in the neurological status and clinical disturbances were not observed.[142]

Changes in EEG and behavioral disturbances were described by McIntire et al. as sequels of severe intoxication in 4 of 31 children poisoned by OCP and diazinon.[143]

Jankins and Toole reported two cases of motor polyneuritis after an occupational exposure to a mixture of aldrin, DDT, and endrin.[144] This effect was attributed to the cyclodienic insecticides, having in mind that DDT-induced polyneuropathies described in the literature were usually motor and sensory. The role of DDT in these two cases was considered to be only secondary.

Chakravarti reported one case of polyneuritis due to DDT and hexachloran exposure in the course of 4 years at a malaria control operation.[145] Precursory symptoms were weakness in the hands and legs and paraesthesia and numbness of the distal part of the extremities. Later, a muscle weakness, difficulty in walking, and ataxia also appeared. A similar case was reported by Schuttmann.[146] A woman with long-term exposure to DDT developed severe polyneuritis.

Onifer and Wisnant also described an intoxication due to inhalatory exposure of DDT and lindane during 3 to 5 months.[147] The effect on the nervous system was characterized by the following symptoms: cerebral ataxia, paraesthesia, impaired motor coordination, areflexia of the extremities, weakness, nystagmus, and ptosis. All symptoms were reversible.

Model and Zaritskaya are of the opinion that the neurovegetative symptoms manifested in chronic intoxications are related to cervical neuritis.[148] They are observed in persons occupationally exposed to either OCP or to other pesticides with similar effects.

Bogusz described tremors in the extremities of workers after a 10-year exposure to OCPs.[149] Frequent symptoms were headaches and giddiness.[150]

Behavioral changes and disturbances of sensory and equilibrium functions have been reported. These symptoms were reversible after discontinuation of exposure.[151]

Loganovski has found changes in the kidney function of persons exposed continuously to DDT, hexachloran, and ethylsulfate.[152]

Krasnyuk studied the cardiovascular function of 400 persons occupationally exposed to DDT and hexachlorane for 1 to 15 years.[153] The cardiovascular morbidity in these subjects was higher in comparison to control subjects. Most often the following symptoms were present: dyspnea, tachycardia, heaviness, heart pains, increased heart volume, and suppressed cardiac sounds. There was systolic murmur in 14.7% of the exposed subjects, against 7.5% in the controls. At X-ray examination, a decreased amplitude of the heart contractions was observed. The ECG revealed heart rhythm disturbances, it showed changes in the duration of the cardiac cycle, duration of the systolic time, and relative frequency of sinus bradycardia in 42 of the cases. The author attributed these disturbances to the direct effect of OCPs, which caused myocardial damage and impairment of the conducting system due to the affected myocardial metabolism. However, Laws et al. did not observe cardiovascular disturbances in workers after a massive exposure to OCP.[154] Later, Krasnyuk published the results of his investigations on 686 subjects exposed to DDT, hexachloran, ethylsulfonate, hexachlorbutadien, polychlorpinen; 39% complained of tachycardia, dyspnea at physical effort, and periodic heart pains.[155] Objective observation established sclerotic and dystrophic changes in the myocardium. It was also found that the dystrophic damages of the myocardium were proportional to the duration of the labor activity.

In the same subjects Boyko described catarrhal, hypertrophic, and vasomotor rhinitis and disturbed olfactory sensibility.[156] Acute toxic myocarditis and tachyarrhythmia were observed in 3 cases of acute poisoning with DDT, dichlorethane, and buthylether.[157] Recovery of the sinus rhythm and improvement of the functional ability of myocardium paralleled the disappearance of intoxication symptoms.

Blood and capillary disturbances, such as anemia, pancytopenia, agranulocytosis, hemolysis, and capillaropathy, also were observed after contact with organochlorine pesticides. In some cases a complete medular aplasia was found.

Hemocapillaropathies (purpura) occur after chronic or acute but massive exposure.

One lethal case of panmyelophtisis resulting from chronic exposure to DDT and BHC was reported by Friberg and Martinsson.[158] Parallel to the prevailing neurological symptoms, they noted a decrease in the red blood cell count, with plasma reticulocytosis and inhibition of erythroblastosis at sternal punction.

A case of pancytopenia with lethal issue was published in 1955.[159] A 16-year-old student in the U.S. received massive exposure to lindane and DDT in the course of two consecutive summer sessions. The patient had a noncontrollable infectious syndrome, pleuritis, intestinal hemorrhages, and necrotic lesions in the pharynx.

Albahary et al. published a similar case of uncontrolled pancytopenia, with prevailing agranulocytosis and moderate liver damage.[160] The case occurred after 3 years of exposure to lindane solution aerosols, preceded by a DDT exposure. It is supposed that in this case the chloramphenicol treatment of the patient had provoked the disease.

Eosinopenia, neutropenia with lymphocytosis, and hypochromic anaemia were established with workers continuously exposed to DDT during its production.[161]

Komarova and Batueva have investigated 365 patients after OCP exposure and found hypochromic anaemia in 29.8% and leucopenia in 6.3%.[162] The authors support that the OCPs accumulated in the organism could provoke the development of some diseases of the hemopoietic system, such as hypoplastic anemia, agranulocytosis, hemorrhagic diathesis, etc.

Hrycek et al. studied 36 workers occupationally exposed to the polychlorinated pesticides DDT and hexachloran.[163] They cytochemically investigated lactate and succinate dehydrogenases in peripheral blood leucocytes. They observed reduced color intensity in neutrophiles for SDH and lymphocytes for both SDH and LDH. The authors suggested that there were indications for disturbances in piruvate metabolism in peripheral blood lymphocytes.

Krasnyuk found more frequent hepatitis and hepatocholecystitis in subjects working with OCPs, while Paramonchik predominantly found functional liver disturbances.[164,165]

Paramonchik and Platonova investigated the liver and stomach function of 70 subjects working with OCPs.[166] In persons with prolonged exposure (more than 10 years), they observed disturbances of the proteinogenic, hydrocarbonic, pigmentogenic, antitoxic, and excretory liver functions, as well as depression of gastric acidogenic function and pepsin secretion and motor disturbances. These findings were connected with asthenia and anorexia. Changes in the secretory function of the stomach have been reported by Platonova.[167] In another study Paramonchik presented data from an investigation of liver functions in a larger group of workers.[168] Paramonchik studied 468 workers (machine operators, packers, repair workers, and laboratory workers) exposed to DDT, hexachloran, and ethylsulphonate during its production. The subjects were up to 40 years old and employed for over 5 years.

The DDT concentrations in the different work places varied from 0.3 to 8.0 mg/m^3 (MAC = 0.1 mg/m^3); hexachloran varied from 0.3 to 8.0 mg/m^3 (MAC = 0.1 mg/m^3) and ethylsulphonate from 0.1 to 3.0 mg/m^3 (MAC = 5.0 mg/m^3). The subjects complained of periodic pains in the right subcostal area, bitterness in the mouth, and dispepsic phenomena, which were more expressed in workers from the hexachloran production and less in those from DDT and ethylsulphate production.

The functional investigations revealed the following data on liver damage, which do not exclude the eventual effect of alcohol abuse in everyday life:

- protein synthesis function: tendency toward increased total protein in blood serum; disturbed ratio between protein fractions and lipo- and glycoproteins in 60% of the subjects (hypoalbuminemia, hypergammaglobulinemia, hypoalphalipoproteinemia, and higher beta-lipoproteins); decreased prothrombin index
- disturbed secretory function (as indicated by the highly sensitive Bengali rose test)[168]
- pigment functions: raised bilirubin in blood
- lipid function: decreased cholesterol in blood

In subjects exposed mainly to OCPs the content of cystin, arginin, serin, glycin, treonin, leucin + isoleucin, etc. were decreased.[169]

Morgan et al. studied the liver functions of workers with high OCP tissue levels.[170] The aim was to check for the presence of liver damage or induction of drug-metabolizing enzymes. The enzymes glutamatoxaltransaminase, glutamatpiruvatransaminase, lactatdehydrogenase, alkaline phosphatase, and creatinin phosphatase were investigated. Delta-glutaric acid is normally found in human urine, but its excretion increases under the influence of endogenic and exogenic substances, because of their effect upon the enzymes involved in its synthesis. It is proved that barbiturates, aminopyrine, phenylbutazone, and contraceptive steroids cause an increase of delta-glutaric acid in animals and human. It has been accepted lately, that its level is an indicator of the microsomal enzyme activity in the liver.

The mean levels of the serum p,p'DDT in the different groups of the subjects investigated were 26, 47, 75, and 225 mg/kg. From all investigated indices some increase was found only in the lactate dehydrogenase which, according to the authors, had no toxicological significance and was only a manifestation of adaptation. No stimulation was found in the excretion of delta-glutaric acid, which was an indication of the absence of microsomal liver enzyme induction.

An indirect indication of the stimulation of microsomal enzymes is the degradation time of some drugs. In workers with OCP exposure (mainly to lindane), phenylbutazone half-life after a single application of 5 mg/kg was found to be 51.1 h compared to 63.9 h in the control group.

Kolmodin et al. studied antipyrin metabolism in 26 men occupationally exposed to chlorinated hydrocarbon insecticides and in a control group.[171] A dose of 10 to 15 mg/kg antipyrin in gelatine capsules was applied. Blood samples were taken in heparinized tubes 3, 6, 9, and 12 h after application. The half-life time of antipyrin was normally distributed, with a mean of 7.7 h (2.7 to 11.7 h). In the control group the mean was 13.1 h (5.2 to 35 h). This study indicated that OCI could induce microsomal drug metabolism in man.

These results were further supported by the study of Poland et al.[172] In people exposed occupationally to high DDT concentrations, they found a stimulating effect of DDT on phenylbutazone metabolism and cortisol hydroxylation. The serum concentration and storage of DDT and related metabolites in adipose tissues were 20 to 30 times higher in the workers compared to the general population. Serum phenylbutazone half-life was reduced by 19% and urinary excretion of 6'-hydroxycortisol was increased by 57%. There was no correlation between the rate of phenylbutazone disappearance and the excretion of 6'-hydroxycortisol in the urine. This suggests that phenylbutazone and cortisol are hydroxylated by different systems that are under separate regulatory control. There was also a lack of correlation between serum DDT levels and the serum half-life of phenylbutazone or the urinary excretion of 6'-hydroxicortisol. The authors explained this finding as genetic differences in individual susceptibility to enzyme induction. Krasnyuk and Platonova stressed on the gastric disturbances observed in 558 workers exposed to OCPs (234 to DDT, 176 to hexachloran and 148 to combined action of different OCPs).[173] Clinical manifestations of gastric pathology have been found in 22.5% of the examined persons who worked with OCPs. The observed gastric pathology was mainly of a functional nature: secretory, enzyme excretion, and motor function disturbances were found. A relationship between the length of exposure in years and the proportion of workers with gastric pathology has been observed. Gastric pathology was found in 14% of the workers exposed up to 5 years, and in 31.8% of those having worked for more than 10 years. The lack of an adequately matched control group hindered the determination of a cause-effect relationship.

Human data on the reproductive effect of OCPs are very limited. o,p'-DDT has been found to have an effect on androgen metabolism. This phenomenon has been used in some attempts to suppress adrenocortical carcinoma.[174,175] Bleherman found a decreased excretion of 17-corticosteroids and 17-oxyketosteroids in women exposed to different OCPs.[136]

Morgan and Roan did not find changes in the adrenocortical function of workers with a long-term occupational exposure to DDT, dieldrin, and other pesticides used in agriculture.[135]

Cannon et al. (cited by Simić and Knieweld) reported oligospermia with predominant abnormal and non motile sperm forms in workers exposed to chlordecone.[176] This effect was accompanied by tremors, nervousness, weight loss, opsoclonia, pruritus, and joint pains. The mean blood chlordecone level in this case was 2.5 ppm. Exposed asymptomatic people had a 0.6 ppm blood chlordecone level. Dieldrin was reported to lead to impotence in spraymen but was successfully treated by methyl testosterone (Espir et al., 1970, cit. by Simić and Knieweld).[176]

Comparisons of maternal blood levels of DDT in cases of normal pregnancy and spontaneous abortion did not show any significant difference.[177] However, Wassermann et al. found that out of 17 cases of premature delivery, 8 were associated with high PCB serum level (28 ppb vs. 18.25 ppb) and 5 were associated with high DDT serum levels (119 ppb vs. 26.5 ppb).[178] In some of these cases high levels of both PCB and DDT were found, as well as high levels of lindane, heptachlor epoxide and and dieldrin. The authors suggest that OCIs may disturb the hormonal balance of pregnancy and thus precipitate delivery. There are several explanations of this effect: estrogenic activity of o,p'DDT and to some extent of other OCIs, MFO induction, and alterations in plasma levels of sex hormones.

Skin irritation through contact with certain OCPs and chlorinated terpens, in particular, is reported. Chronic intoxications are often clinically demonstrated by signs of allergic damage.

1. Effects of Individual Compounds

DDT mainly affects the nervous system (central and peripheral) causing hyperexcitability, tremors, ataxia, epilepsia, seizures, and convulsions.

Acute intoxications with DDT are less frequent, and often they are due to negligence or ignorance.

Pushkar and Bryazgunov described intoxication of a mother and her two children ages seven and one, after they consumed strawberries and red currants treated with DDT.[179] Part of the fruits were eaten unwashed and fresh, and the rest — boiled as compote fruits. Analysis of the residue of the compote fruits established the presence of DDT. Twenty-four hours after the fruit intake, all three patients were in a state of severe intoxication — temperatures up to 40.1°C, vomiting, and diarrhea. The patients were hospitalized. After clonictonic convulsions the 7-year-old girl lapsed into unconsciousness and died. The other child was in a grave condition, as well, but after the fourth day it improved; on the 15th day the 1-year-old was discharged from hospital. The authors concluded that DDT affects mainly the CNS and contributes to the development of hypertermia and cramps. In humans, some subjective symptoms, such as prickly sensations in the tongue and around the mouth and nose, are also experienced. Reduction of the tactile sense, paraesthesia of the extremities, dizziness, confusion, headaches, malaise, and restlessness may also occur.

Acute occupational intoxication is also reported by Conley.[180] After 1 year of DDT exposure, the worker noticed bleeding gums, a sore throat, and red spots on his tongue and over the rest of his body. Completely aplastic bone marrow was revealed. One week later he died from a massive hemorrhage, secondary to thrombocytopenia.

Kolarski described two intoxications with DDT contained in baked flour.[181] In both cases, cramps were followed by full recovery.

A single DDT dose of 10 mg/kg produces illness in some subjects, 16 mg/kg causes convulsions. The estimated lethal dose is 500 mg/kg, but when kerosine is used as a solvent it is 150 mg/kg.[182]

Death is usually due to respiratory failure at the convulsive stage of poisoning or to ventricular fibrillation. The mechanism of DDT intoxication appears associated with its effects on the membranes in the nervous system, causing changes in the movement of both Na^+ and K^+ through the axonal membrane. Other evidence of nervous system effects are changes in the concentrations of norepinephrine and other neurotransmitters.[183]

Chronic DDT occupational poisonings are characterized predominantly by damage to the nervous system.

In 123 workers with long-term exposure to DDT (from 1 to 10 years), Model observed disturbances in the peripheral nervous system, polyneuritis, and polyneuromyalgia in upper extremities.[183,184] To these disturbances were added asthenic state, hypohondria, vasomotor and vestibular crises, apathy, and memory weakness.

A case of polyneuritis due to DDT was noted by Hermann.[185] It concerned a hospital employee, engaged in disinfection, who worked with DDT for a number of years. After having changed his job for about 9 months he began to work with DDT again. Digestive and renal symptoms then appeared, and he later developed some neurological disturbances like the disappearance of the knee reflex and an incapacity for walking. Besides that, muscle atrophy and paraesthetic disturbances were noted. The polyneuritis had a favorable development, and within 6 months the patient completely recovered. However, it was difficult to establish to what extent DDT caused this polyneuritis, since no toxic level was found in subcutaneous adipose tissue.

Another case of polyneuritis of the upper extremities, after an accidental intake of DDT, was reported by Michon et al.[186] A single dose of 100 g of 5% DDT powder mixed with water was ingested. Symptoms of acute intoxication appeared after 20 h with a hypotonia of the upper extremities, disappearance of knee and Achyles reflex, paraesthesia, diffuse disturbances of surface sensitivity, etc. Complete recovery occurred in 2 weeks.

Neurological symptoms of workers engaged in DDT production were observed by Alexieva et al.[187]

It is well known that DDT provokes tremors in rest, its intensity being dependent on the concentration of the chemical in nervous tissue. This DDT-induced tremor did not depend on the extrapyramidal system and persisted more than 24 h.

In India, Misra et al. observed 29 workers who had sprayed 5% DDT for a long period of time (1 to 32 years) without protective clothing.[188] They attempted to quantify the effects on the cognitive functions as sensitive indicators of neurotoxicity. Bhatia Battery Performance Test of Intelligence, Wechsler Memory Scale and Bender Visuomotor Gestalt Test (BGT) were used. The EEGs of 9 subjects who had abnormal BGT performance were obtained. To assess the degree of exposure to DDT, its isomers and metabolites were estimated in 17 sprayers, and 12 matched controls. Serum levels of the total DDT in workers was 8.5 times higher than that of the controls (0.401 ppm against 0.047 ppm), and p,p'DDT was 31 times higher, which pointed to a recent exposure. The symptoms were mild and did not alter worker activity: headaches in 37.9%, ocular symptoms in 20.7%, and rawness in throat in 17.2%. Some psychological symptoms were reported by 27.5% (irritability, insomnia, depression, asthenia, agitation, and forgetfulness), but only 10.3% of the workers showed worsening symptoms after spraying.

Physical neurological examination revealed moderate signs in 24% of the sprayers — generalized hyperreflexia and deep tendon jerks, tremulousness, and loss of ankle jerks and fasciculations. Since no control group was used for comparison, their relationship with DDT was not clear.

The intelligence and memory quotients did not show significant difference compared to controls. BGT scores of the workers were higher and showed moderately significant correlation with DDT levels in serum. This is interpreted by the authors as the involvement of a dominant frontoparietal lobe. EEG changes were present in 55.5% of the sprayers, alpha activity was absent in 4 subjects, and 1 subject showed diffuse slowing and asymetry. These results were consistent with diffuse involvement. The authors believed that in the light of their study, the potential role of DDT in the etiology of mental retardation, minimum brain disfunction, and presenile demention needs to be carefully studied.

One of the most severe toxic effects of DDT is liver damage.[189] In a case with lethal issue centrolobular necrosis was established.[190] Carlson and Kolmodin-Headman found hyperlipidemia in workers exposed to DDT.[191] They found hyperalphalipoproteinemia in 40% of the workers — hypercholesterinemia and hyperglyceridemia in a lesser number.

Full-time occupational exposure stimulated microsomal enzymes of the liver, where 25 mg/kg or more is absorbed. The same degree caused an increased incidence of tumors in male mice of a susceptible strain.

No epidemiological evidence of such effects in humans has been observed. Elevation of AP and serum glutamic transaminase is found in workers exposed to DDT.

The cardiovascular system is more often affected. Palpitations, tachycardia, and irregular heart action have been noted in some cases of acute poisoning.

The data on hemopoietic damage as a result of DDT exposure exist. Wright et al. reported one case of agranulocytosis, possibly caused by exposure to DDT and dichlorfluormethane.[192] It concerned a student, who had desinsected his home with an aerosol bomb. The symptoms appeared 10 d after the exposure and included a higher temperature, sore tongue, and lymphadenopathy.

Higher values of p,p'DDT were found in the blood of patients with certain hematological syndromes (pancytopenia and thrombocytopenia) compared with the control group.[193] Karpinski reported four cases of purpura in small children under three years of age.[194] The manifestations of purpura were very well expressed and connected with thrombocytopenia in all four cases. After symptomatic treatment and environmental change, all symptoms disappeared quickly. The author considered the contact with DDT used for home desinsection as a possible cause.

A case of hemorrhagic purpura after one week contact with DDT is reported by Rapport of the Council of Pharmacy and Chemistry (cited after Albahary et al.[160]).

A rubella-like rash on a child who had had contact with DDT was a possible result of intoxication.[195] DDT determination in biological media has shown p,p'DDE in blood — 0.18 mg/kg and in urine — 0.03 mg/kg. In a case of acute suicidal oral intoxication of DDT, a post mortem analysis revealed 0.8 mg DDT per 100 g liver and 4.2 mg DDT per 100 g stomach tissue.[180]

The damage to the kidney functions of workers exposed to DDT in workers for more than 10 years was reported.[196,155]

DDT provokes skin reactions, especially after skin contamination. In persons who have had skin contact with DDT solution in kerosine sprayed on blanket and linen, the first symptoms appeared after 24 h: true burns of first degree on the back and backsurface of the thighs, accompanied by very strong itching and headache. Some were less affected, and one had such symptoms as erythema, itching, small eczematous vesicules, and a slight headache. The role of kerosine was also taken into consideration in this case.

One of the metabolites of DDT-o,p'DDT, affects adrenal secretion and extraadrenal metabolism of cortisol.[176,198] It was noted that the extra adrenal metabolism of cortisol was altered so as to decrease the proportion of cortisol excreted as tetrahydrocortisol and tetrahydrocortisone and to increase the proportion excreted as 6'-hydroxycortisol.[198]

The clinical picture of chronic intoxications with hexachlor-cyclohexan (HCH) in humans has yet to be completely elucidated. Burkatskaja et al. presented the health data of workers engaged in hexachloran production.[200] The publications of Kalyaganov,[201] Odintzova,[202] and Gorskaya et al.[203] draw attention to the effects on the nervous and cardiovascular systems, the liver, the pancreas, and blood. The concentrations of hexachloran in the working premises were found to be higher than the MAC value. The workers were divided into two groups depending on the duration of their working activity: first group — with 11 to 15 years of work activity (24 workers) and second group — with more than 16 years (36 workers).

The most frequent complaints were headaches, fatigability, dispnea at physical effort, irritability, pains in the heart area, stinging in the epigastrium, itching, and eruptions on the open body surfaces. An examination of the cardiovascular system revealed dull heart tones and a systolic murmur on the top with functional character. The ECG showed sinus bradycardia, in some cases sinus arhythmia, disturbed ventricular auricle conductance, and dystrophic myocardial changes in 97% of the cases. Hainz bodies in erythrocytes and eosinopenia were also found.

Chronic hepatitis was diagnosed in 39%, bile ways dyskynesia in 55%, and chronic pancreatitis in 6% of the subjects investigated. In the course of the 2 year dynamic observation, the chronic hepatitis was manifested by dyspeptic symptoms, pain, and astheno-vegetative syndrome. At clinical examination, hepatomegalia and, in single cases, splenomegalia was found. Cholecystography revealed gall bladder disturbances. Biochemical tests for assessing the liver function have shown increased SGTP activity, decreased AP and ChEA, hypoalbuminemia, and lower prothrombin index.

The external secretory function of the pancreatic gland, studied in 25 patients, has shown decreased lipase, tripsin, and amylase activity in 80% of the cases. The disturbances of the external secretory function have been of a hyposecretory type, with signs of dyspancreatism, and proceed without characteristic clinical manifestations. In 80% of the cases the glycemic curves have been pathological — flat or of an irritative type.

Long-term exposure to hexachlorane induces chronic hepatitis, which usually inhibits the extrasecretory function of the pancreas, without clinical symptoms. Parallelly, a chronic pancreatitis of a latent development is observed.

A liver-kidney syndrome, provoked by exposure to lindane, is described by Wakatsuki.[205]

An investigation on 184 desinsectors working with HCH found clinical manifestations in 45.9%. The symptoms disappeared in 7 to 8 months after discontinuing work. Neurological effects (headaches, dizziness, loss of balance, nausea, and vomiting) were found in 79% of the cases, dermatological symptoms (inflammatory dermatitis and skin allergy) — in 20%, gastrointestinal (dyspeptic) — in 11.7%, and inhalatory (chemical pneumopathies, and respiratory allergy) — in 16%. These manifestations did not influence working capacity. Laboratory studies showed hypochromic anemia, increased alpha- and gamma-globulins, and disturbed liver detoxication. For a period of 9 years, 17 cases of acute intoxication by HCH have been registered.[206]

Severe disturbances of hemopoiesis, provoked by HCH and lindane, have been described.

A case of agranulocytosis was reported by Marchand et al.[207] In an environment in which hexachlorcyclohexane was evaporated by an electrical fumigator several hours daily, a worker developed agranulocytosis. The several-month exposure caused the disappearance of the granulous line simultaneously in blood and bone marrow. The patient recovered quickly after discontinued exposure. The authors discuss the possibile sensibilizing action of HCH; the agranulocytosis could be a reaction of intolerance toward this agent.

Two cases of HCH-induced aplastic anemia were reported by McLean.[208] Three other cases, related to the combined effect of DDT and lindane, were reported by Woodliff et al.[209]

Loge presented data for two cases of aplastic anemia, possibly caused by continuous contact with lindane.[210] One of the cases concerned an 8-year-old girl who had been spending 2 h daily, over the course of 2 years, in a home contaminated with lindane. One year after the appearance of echymoses the girl died. A concentration of 0.4 mg/kg lindane was found in her liver (against 0.1 mg/kg, approximately, in the general population). The second case was a 52-year-old man who worked in contact with lindane over the course of several months; he died from pneumonia after pancytopenia was diagnosed.

Disturbances in the nervous system have been also described. Tolot et al. published data on a case of sensomotor polyneuritis due to lindane.[211] It concerned a farmer who had worked with great quantities of lindane mixed with a solvent 15 d before the appearance of symptoms.

Kalyaganov reported eruptions on the open body parts in workers exposed to hexachlorcyclohexane.[201] Data on severe HCH poisoning are reported by Danopoulos et al. as well.[213]

In everyday life poisonings occur mainly with the α-isomer of HCH, lindane.

Starr and Clifford described a case of acute lindane poisoning in a $2^1/_2$-year-old child.[214] After the intake of several lindane granules, the intoxication developed violently. Lindane and its metabolites were determined in the blood, urine, and feces. After treatment, the condition improved and the child was discharged from hospital.

Bambov et al. reported a group intoxication; 11 persons consumed lindane mixed with the sugar in coffee.[215] In spite of the severity of the clinical symptoms (loss of consciousness and cramps), all the subjects recovered after 24 h of active therapy. The intake per person was calculated to be 0.6 mg/kg.

Osuntokun also described cases of lindane poisoning.[216] The characteristic symptoms were bloody vomiting, generalized cramps, and loss of consciousness. Treatment resulted in recovery without sequels.

Kay et al. reported lethal poisonings of persons who had consumed lindane-contaminated meat.[217] The lindane intake was 0.3 g/kg (for comparison, LD_{50} for rats is 0.2 g/kg).

The presence of lindane in blood is an indicator of exposure. Changes in the EEG, a rise of blood pressure, and a fall in heart rate are often related to lindane intoxication.[31]

Kashyap et al. studied the effect on malaria spraymen exposed to HCH.[218] They found significantly higher concentrations of total HCH residues in all exposed groups of spraymen, compared to the controls: 286 to 595 µg/l vs. 174 µg/l, respectively. In the follow-up study, the amount of residues found was 5 times higher in subjects engaged in spray operations for the first time as compared to those who had been previously exposed, probably as a result of enhanced metabolism due to enzyme induction. The significant increase in blood sugar suggested neoglycogenesis or insulin depression, according to the authors. Increase in al/gl ratio without specific change in total protein indicated suppression of globulin synthesis. These changes were easily reversible.

2. Cyclodiene Compounds

The cyclodiene compounds aldrin, dieldrin, endrin, etc. are among the high toxicity substances within the OCP group.

In a group of 297 workers participating in filariasis and malaria prevention and control, about 20 dieldrin intoxications were observed. The poisoning began with slight muscle twitches; they developed further with recurrent convulsions (several times daily) and a loss of consciousness for several hours. In some cases the convulsions appeared later, and in one subject, the first attack was observed 15 d after cessation of work. Along with the twitches symptoms such as headaches, nausea, general malaise, and dizziness were observed.

A case of lethal issue after several hours of occupational exposure has been reported by Symansky.[220]

Nervous system disturbances are often observed in intoxications by aldrin and eldrin during their production. In 15 workers with convulsions, Hoogendam et al. registered EEG abnormalities which persisted after seeming recovery.[221,222] Periodic examinations of the workers in this factory occasionally have shown the same EEG abnormalities without any clinical syndrome being present. These abnormalities have been considered early symptoms. Workers occupationally exposed to dieldrin had blood levels of 0.15 to 0.44 µg/ml. At the level of 0.15 µg dieldrin/ml, neither health impairment nor signs of enzyme induction were observed.[4]

Complete and continuous amnesia, 8 months after severe intoxication with dieldrin, has been reported by Jacob and Lurie.[223]

Behavioral disturbances have been observed in experiments on cats, poisoned by 1 mg/kg daily dieldrin dose after 13 applications.[224] In monkeys, behavioral disturbances were induced by aldrin at doses no higher than the daily quantities absorbed by some groups for a period of several weeks (Thomson, cited after Mrak[27]).

According to Pick et al. a case of aplastic anemia resulted from recurrent contact with aldrin.[225] After 6 months of pesticide spraying (aldrin, parathion, and metaldehyde) the person developed intoxication. After being treated for 1 month, the subject returned to work, but a subcutaneous hemorrhage appeared; 5 d later he died from bleeding in the brain.

Kazantzis et al. observed 10 workers who worked for 1 week at high concentrations of aldrin.[226] In four of them, EEG disturbances (low voltage and irregular rhythm) and some vision symptoms were found, four had vomiting, and three had loss of consciousness.

Coble et al. reported three cases of endrin poisoning with cramps.[227] The following concentrations of the toxic substance were determined: 0.021 to 0.093 mg/dm³ in serum and 0.02 to 0.039 mg/dm³ in urine. All three cases improved after treatment.

Preda et al. described the acute intoxications from bread prepared with endrin mixed flour, and unwashed grapes treated with aldrin.[228] Nausea, vomiting, tenesmus, excitation, hyperreflexia, irritability and coma appeared very quickly — only 3 min after eating. Three of the subjects died from brain edema and toxic hepatitis. One of the survivors showed signs of euphoria and complete amnesia.

Fokim investigated a group of women who had an acute poisoning by polychlorcamphen.[229] The first symptoms were observed 1 to 5 h after exposure; they manifested themselves as irritation of the eyes and upper respiratory ways, ardor in the epigastrium, dryness in the mouth and throat, headaches and dizziness, nausea and frequent vomiting, severe muscle weakness, paraesthesia in the lower extremities and generalized tremor, and loss of consciousness for 15 to 20 min. On the eighth day, from the beginning of the clinical manifestations, anisocoria, anisoreflexia, myosis, and decreased muscle tonus and strength were objectively established. CNS functional changes, astheno-vegetative syndrome, and vegetative vascular dystonia were the leading forms of mild intoxication. For the moderate and severe forms of poisoning, encephalopolyneuritis and diencephalic syndromes and transitory disturbances of brain circulations were characteristic.

Warraki reported two occupational toxaphene poisonings after a several month exposure.[230] At clinical investigation, the acute bronchiolitis and convulsive crises were prevailing. Radiological examination of the lungs revealed bilateral lymphadenopathy, miliary shadows of both lung fields, most expressed in the lower part. Blood analysis showed light leucocytosis, eosinophylia, and increased globulins. The author diagnosed acute allergic pneumonia.

The estimated minimum acute lethal oral dose of camphechlor for man is reported to be 2 to 7 g; doses of 10 mg/kg or less provoke nonfatal convulsions in some cases but are not effective in others. Dermal single application of 46 g is very dangerous. Inhalatory exposure of volunteers to camphechlor at a concentration of 500 mg/m³, 30 min/d for 10 d failed to produce any adverse effect.

A family was poisoned by eating collards recently sprayed with 9% camphechlor. Of 10 persons, 7 developed symptoms. Doses of 9.5 to 45 mg/kg of camphechlor were estimated to have been consumed.[231]

Garretson reported the accidental intoxication of two children who ingested dieldrin.[232] In one, the symptoms of poisoning appeared 15 min after intake; a half an hour later the child died. Heart dilatation and stasis in the lungs were found post mortem. In the other child, who recovered after three days treatment, convulsions, cyanosis, and 39.5°C body temperature were observed. On the third day, the serum level of aldrin was 0.27 mg/dm³, 50 times the respective value for the general population; the concentration in the adipose tissue was 47 mg/kg against 0.29 mg/kg in general population and 6.12 mg/kg in occupationally exposed workers. Alkaline phosphatase and coagulation tests still showed increased values 6 months later.

Van Raalte reviewed all lethal intoxications with endrin, dieldrin, and aldrin.[233] From a total of 97 intoxications with endrin, 69 were suicides and 4 were occupational poisonings. A dose of 10 mg/kg dieldrin was reported to be toxic for man. Convulsions in children were observed at a dose of 44 mg/kg.[4]

The clinical picture of chlordane poisoning is similar to that of the other cyclodiene compounds. Aldrich and Holmes reported a case of acute chlordane poisoning in a child.[234] Cramps and myoclonia appeared 3 h after the intake of an undetermined quantity of chlordane. Other symptoms of the poisoning were hyporeflexia, loss of coordination, leucocytosis, and strengthened outline of the vessels and bronchi at radiological examination. Serum concentrations of chlordane on the 1st, 3rd, and 13th day of intoxication were respectively 3.4, 0.138, and 0.03 mg/dm³. The calculated half-life of chlordane in serum was 88 d. Chlordane levels in the urine were 0.05 mg/dm³ and 0.13 mg/dm³ on the 3rd and 14th day respectively.

In a patient poisoned by chlordane, Curley and Gerretson found, just after the cramps, 2.71 mg/kg chlordane in blood and 3.12 mg/kg in adipose tissue.[235] The half-life of chlordane in serum was calculated to be 21 d. The blood/adipose tissue ratio of chlordane was variable. It was 10 times greater 3 months after the intoxication — 0.11 in blood and 25.5 mg/kg in tissues. Deviations in serum alkaline phosphatase and coagulation test persisted 2 months after the intoxication.

In a chlordane poisoning described by Stranger and Kerridge, the severe cramps caused multiple vertebral fractures.[236]

Fishbein et al. did not find any adverse health effect in workers engaged in its chlordane production.[237] With exposure ranging from 1 to 15 years, the concentration of chlordane was fairly constant, with levels from 1.2 to 1.7 mg/dm^3 in the working zone.

It has been estimated that the minimum dermal dose of heptachlor required to produce symptoms in man is 46 g for a single dose and 1.2 g for repeated exposure.[231]

In a study on agricultural workers exposed to polychlorpinen (PCP), disturbances in the cardiovascular system have been found.[157] Some authors explain these changes with an immediate effect of PCP upon myocardium, as well as with the increased catecholamine quantity.[238] Functional neuromuscular disturbances and changes in the rheobase of the muscles-antagonists have been reported.[239] Besuglij et al. have found impaired liver secretory function, increased urobilirubin excretion, decreased cholesterine, and β-lipoproteins.[240] Changes have also been found in the protein synthesis of the liver — hypoalbuminemia with hypergammaglobulinemia.[241] Disturbances in aminoacid metabolism, e.g. increase of serin, aspargin, and alanin content and decrease of glutamic acid, have been observed.[242] Evidence to the polytropic effect are the changes in the visual and vestibular function,[240] disturbances in the ovarial menstrual cycle, etc.[243]

Shapovalov described cases of severe acute intoxication with methylbromide; the two women did not observe the safe work regulations.[244] One of the patients died of toxic pulmonal edema and coma, and the other survived after coma and bromine psychosis. After a complex treatment, she was discharged from the hospital in good health condition.

Chlordinitrobenzene is one of the strongest skin irritants. If used as an algicide it causes skin damage; it is recommended only in closed systems.[245]

Among the other OCI, hexachlorbenzene (HCB) should be noted with the peculiarities of its action. It is used for seeds dressing, usually as 30% dust. Due to low toxicity, only chronic intoxication is known.

During an epidemic poisoning in Turkey from HCB-treated grain, photosensibilization, liver injury, and porphyria were manifested. A detailed description of the epidemy is given by Schmid.[246] More than 4000 people were intoxicated over a period of 5 years, from 1955 to 1959. The estimated intake of HCB was 50 to 2000 mg/d for a long period.

The intoxication was characterized by photosensitive blistering and epidermolysis on the skin of the hands and face. The skin becomes unusually sensitive both to light and minor mechanical trauma. Blisters break easily, form crusts, and get replaced by pigmented scars containing microcysts. Scarring leads to contractures, permanent alopecia, or corneal opacity. Additional infection may develop into arthritis and osteomyelitis. Pigmentation and fine dark hair appear around the eyes and chin and on the extremities (monkey disease). Colic and severe weakness, to the extent that the patient complains of being unable to handle eating utensils have been reported, as well as enlarged thyroid, subnormal temperature, weight loss, and muscle atrophy.

HCB is secreted in mother's milk and produces intoxication in children. HCB is a teratogen; liver enlargement and functional disturbances have often been noted. Porphyrin has been found in urine; in such patients urine is portwine red and dark. HCB-induced rise of cytochrome P-450 content has been experimentally established, as well as an increase of the *in vitro* activity of some oxidative microsomal enzymes and the *in vivo* decrease of the pharmacological effect of hexabarbital.[247]

According to the Schmid study, despite ceasing the consumption of contaminated bread, the disease persisted with further complications, such as arthritic joint changes, hand contractures, osteomyelitis, etc.[246] At in-patient treatment an improvement was observed, but not a disappearance of the illness; the mortality rate was 14%. It was suggested that insufficient food protein and undernutrition enhanced the development of disease. Children, born by mothers who had often eaten the grain, developed pink sores with 50% mortality.

Gocmen et al. performed a follow up study on 204 patients with a history of HCB-induced porphyria during their childhood.[248] After some 25 to 30 years, many symptoms were still persisting, such as scarring of the face and hands (86%), hyperpigmentation (71%), hirsutism (47%), pinched faces, fragile skin (37%), painless arthritis (66%), small hands (64%), enlarged livers (4.4%), and enlarged thyroids (37%). Seventeen persons had higher levels of one of the porphyrins, and eight were considered still porphyric. The average value of HCB in breast milk was 0.51 ppm (0.07 in the controls). Elevation of the delta-aminolevulinic acid in over 50% of the patients was present.[249] There was a significant evidence of sensorymotor neuropathy and myotonia and "cogwheeling", considered as an onset of extrapyramidal symptoms.[257,258] Thyroid enlargement was associated with a slight reduction of the thyroid function. A positive effect of agents (ethylenediaminetetracetic acid) in the acute phase was demonstrated.[250,251] Stimulation of fecal HCB excretion using paraffin and mineral oil or other lipophilic compounds (squalane) was reported.[252]

The second outbreak of HCB poisoning occured in Saudi Arabia. The source was a bakery using HCH for deratting. The bread was baked with 220 ppm. About 450 persons were hospitalized, 7 died.[253]

3. Treatment[39]

Cases of acute intoxication should be treated as follows:

1. Remove the poison by washing the skin with soap and water. Stomach lavage is made, first with charcoal, then with salt-based cathartics.
2. No specific antidote for OCl poisoning has been found. Treatment is aimed at controlling the symptoms, especially hyperactivity, tremors, and convulsions. Treat with soluble barbiturates and diazepam in sufficient doses to calm the patient and prevent convulsions. Give i.m. injections of phenobarbital in doses 0.1 to 0.2 g. In cases of violent convulsions, i.v. administration of active barbiturates (such as 0.25 to 0.5 g of phenobarbital) is indicated.
3. Administer 3000 to 5000 ml glucose, calcium gluconate, and saline solution, taking care to check the hemodynamic parameters.
4. Use supportive treatment to maintain respiration and oxygenation and to prevent secondary infection, when indicated.
5. Adapt specific therapy to the symptoms to protect the liver (hepatoprotective therapy by corticosteroids and levulose solutions). Take preventive measures against kidney failure, such as dyalisis.
6. Hospitalize any patient who suffers from acute intoxication with convulsions for at least 1 week.
7. Administer epinephrine, it sensitizes the heart to abnormal rhythms.

Chronic intoxication is treated symptomatically: 20% glucose solution (up to 1000 ml twice daily), 1000 mg Vitamin C twice daily; 10% calcium gluconate solution (10 ml i.v. 2 to 3 times daily), anti-allergic treatment, and rehabilitation procedures.

The chances of complete recovery are good. In heavy intoxications with prolonged convulsions, there is a possibility of permanent brain damage due to continued anoxia.

Treatment of HCB intoxication differs from that of other OCIs. It includes gastric lavage, using 5% sodium bicarbonate and washing the skin with soap. The patient should be kept away

TABLE 13
Maximum Permissible Levels of OCI in Air of Working Environment
(in mg/m^3)[254]

Pesticide	U.S.	U.S.S.R.	WHO	IDLH[a]
Aldrin	0.25	0.01		Ca[b]
DDT	1	0.1		Ca
Dieldrin	0.25	0.01		Ca
Endrin	0.1			200 mg/m^3
Hexachlorcyclohexane		0.1		200 mg/m^3
Lindane	0.5	0.05	0.3	1000 mg/m^3
Hexachlorbenzene		0.9		
Heptachlor	0.5	0.01		100 mg/m^3
Chlordane	0.5	0.01		500 mg/m^3

[a]   Immediately dangerous to life or health level.
[b]   Ca — potential human carcinogen.[256]

from direct sunlight. Chelating agents have been used with some success: EDTA (disodium salt) administered i.v. as 1.5 g EDTA in 1000 ml 5% glucose in water daily for 5 d. This treatment is followed by disodium EDTA given orally as 1.5 g daily for 16 weeks, 1.0 g for 10 weeks, 0.5 g for 8 weeks, 0 for 4 weeks, and then 1.0 g for another 10 weeks.

Prognosis is poor in severe cases. Photosensibility may reappear up to 5 years after exposure. Complete recovery occurs only in light intoxications.

4. Prevention

The health hazards of organochlorine pesticides are present at all stages of their lifecycle. During production, acute intoxications are very rare, and chronic intoxications are not a serious problem. When good occupational hygiene practice is observed, even long-term (11 to 15 years) work in DDT production should not produce any ill health effects attributed to exposure.[154] Maximum permissible levels should be followed (Table 13). In agriculture, where large numbers of low skilled workers are engaged and efficient collective measures are difficult if not impossible to maintain, the potential risk for intoxications is considerably increased.

Good technical training and the proper operating conditions of the equipment are necessary. Pre-employment medical supervision and periodic medical examinations should be designed to determine who should not work with OCIs, for example, persons with diseases of the nervous system, liver, or blood, individual susceptibility, or any signs and symptoms of early intoxication. For biological monitoring, safe levels in blood are shown in Table 14.

Most OCIs are skin irritants, some with high dermal toxicity. Therefore, the skin should be well protected from contamination through the proper use of protective clothing and gloves. Where indicated, respiratory protective means should be used. The OCIs are stable, and a thorough daily cleaning of clothes and personal protective equipment should be organized by the employer.

The general population can be exposed to OCIs mainly through food containing OCI residues. Acceptable daily intake values are recommended by WHO/FAO (Table 15).

VIII. CONCLUSION

The group of organochlorine pesticides (OCPs) covers a wide range of compounds, from the cyclodiene series, halogenated aromatic compounds, cycloparaffins, chlorinated terpens, etc. They are used as insecticides, acaricides, fungicides, and rodenticides. Their high persistence in the environment is one of the reasons for their decreased use in many countries, especially DDT which was most used in the past.

TABLE 14
Biological Safe Levels in Whole Blood in Human[138,255]

Pesticide	Individual maximum safe level		References
Aldrin and dieldrin	0.1	µg/ml	138
DDT	0.2	µg/ml	138
Lindane	0.02	mg/l	255

TABLE 15
Acceptable Daily Intake of Pesticides for Man (in mg/kg/b.w./d)[254]

Aldrin	0.0001
DDT	0.005
Endrin	0.0002
Dieldrin	0.0001
Lindane	0.01
Metoxichlor	0.1
Heptachlor	0.0005
HCB	0.0006
Chlordane	0.001

Acute oral OCP toxicity differs, depending on the chemical structure of the compound. Thus, LD_{50} varies from 7 (endrin) to 800 mg/kg/b.w. (keltan and methoxychlor).

OCPs induce microsomal enzyme production and increase the smooth endoplasmic reticulum. As a sequence of this process, interference with steroid and drug metabolism may occur.

OCPs accumulate in the adipose and other tissues. In the adipose tissues of the general population the average values for the different organochlorine pesticides are DDT total — from 1.75 to 30, HCH total — from 0.16 to 2.43, dieldrin — from 0.046 to 0.68 and heptachlorepoxide — from 0.0085 to 0.19 mg/kg. Maximum OCP concentrations in the fat tissues of exposed workers are 10 to 100 times higher.

OCPs are excreted through human milk. Biological monitoring organized by UNEP/WHO demonstrated the following mean values in milk fat: p,p'DDE — 0.81 to 4.4, HCH — 0.05 to 6.6, and PCBs — 0.45 to 2.1 mg/kg. Sucklings in some countries are exposed to OCP levels which sometimes are several times higher than the allowable daily dose.

Acute OCP intoxications do not represent a serious problem, except for aldrin, eldrin, dieldrin, and toxaphene. Chronic intoxications are more common.

The nervous system is the principal target for OCPs. Polyneuritis and encephalopolyneuritis are reported after occupational exposure to OCPs. Tremors and convulsions are typical symptoms for OCP effect.

Blood and capillary disturbances are often related to OCP exposure. The liver and gastrointestinal system are also often affected: chronic hepatitis or different grade of function disturbances.

Skin irritation and allergic reactions by contact with certain OCP are reported.

There is no specific antidote therapy for OCP intoxications. The treatment should be symptomatic and complex.

REFERENCES

1. **Edwards, O.,** *Persistent Pesticides in the Environment,* Butterworths, London, 1970, 67.
2. DDT and its Derivatives, Environmental Health Criteria document, No. 9., World Health Organization, Geneva, 1979.
3. **Feldmann, R. J. and Maibach, H. T.,** Percutaneous penetration of some pesticides and herbicides in man, *Toxicol. Appl. Pharmacol.,* 28, 126, 1974.
4. Dieldrin, Data Sheet on Pesticides, No. 17, World Health Organization/Food and Agriculture Organization, unpublished document, 1975.
5. **Uhi, S., Schmid, P., and Schlatter, C.,** Pharmacokinetics of pentachlorphenol in man, *Arch. Toxicol.,* 58, 3, 182, 1986.
6. *Pesticides Dictionary,* Willoughly and Meister, 1978.
7. **Conney, A., Welch, R., Kuntzman, R., and Burna, J.,** Effect of pesticides on drug and steroid metabolism, *Clin. Pharmacol. Therap.,* 8, 2, 1966.
8. **Kuntzman, R., Sansur, M., and Conney, A.,** Effect of drugs and insecticides on the anaesthetic action of steroids, *Endocrin.,* 77, 952, 1965.
9. **Kupfer, D.,** Effects of some pesticides and related compounds on steroid function and metabolism, *Res. Rev.,* 19, 11, 1967.
10. **Hart, L. and Fouts, J.,** Studies on the possible mechanism by which chlordane stimulates microsomal drug metabolism in rat, *Biochem. Pharmacol.,* 14, 26, 1965.
11. **Hart, L. and Fouts, J.,** Effects of acute and chronic DDT administration on hepatic microsoma drug metabolism in the rat, *Proc. Soc. Exp. Biol. Med.,* 114, 388, 1963.
12. **Davies, J., Edmunson, W., Schneider, H., and Cassaday, J.,** Problems of prevalence of pesticides residues in humans, *J. Pest. Monitoring,* 2, 80, 1968.
13. **Davies, J. and Edmundson, W.,** The pros and cons of depesticidation in man, *Ind. Med. Surg.,* 39, 7, 320, 1970.
14. **Rappolt, R.,** Use of oral DDT in three human barbiturate intoxicaitons: CNS arousal and/or hepatic enzyme induction by reciprocal detoxicants, *Ind. Med. and Surgery,* 39, 7, 319, 1970.
15. **Schoor, W. P.,** Effect of anticonvulsant drugs on insecticide residues, *Lancet,* ii, 520, 1970.
16. **Kwalik, D. S.,** Anticonvulsants and DDT residues, *J. Am. Med. Assoc.,* 215, 120, 1971.
17. **Watson, M. Gabica, J., and Benson, W. W.,** Serum organochlorine pesticides in mentally retarded patients on differing drug regimens, *Clin. Pharmac. Ther.,* 13, 186, 1972.
18. **Sizonenko, P. C., Doret, A. M., Riondel, A. M., and Paunier, L.,** Cushing syndrome due to bilateral adrenal cortical hyperplasia in a 13-year-old girl: successful treatment with o,p'-DDT, *Helv. Paediatr. Acta,* 29, 195, 1974.
19. **Murakami, M. and Fukami, Y.,** Incorporation of labeled pesticides and enviromental chemicals into nuclear fraction of cultured human cells, *Bull. Environ. Contam. Toxicol.,* 24, 1, 27, 1980.
20. **Wassermann, M., Wassermann, D., Gershon, Z., and Zeller-Mayer, L.,** Effects of organochlorine pesticides on body defense system. Biological effect of pesticides in Mammalian system, *Am. N.Y. Acad. Sci.,* 160, 393, 1969.
21. **Fitzhug, O. and Nelson, A.,** Chronic oral toxicity of DDT, *J. Pharmacol. Exp. Ther.,* 8, 84, 18, 1947.
22. **Davies, K. and Fitzhugh, O.,** Tumorigenic potential of aldrin and dieldrin for mice, *Toxicol. Appl. Pharmacol.,* 4, 187, 1962.
23. **Kemeny, T. and Tarjan, R.,** Investigations on the effects of chronically administered small amounts of DDT in mice, *Experim.,* 22, 748, 1966.
24. **Tomatis, L.,** Studies on the potential carcinogenic hazard presented by DDT, in Whither Rural Medicine, Proc. 4th Int. Congr. Rural Med., Usuda, 1969 (Tokyo), 1970, 64.
25. **Tomatis, L.,** The effect of long-term exposure to DDT on CF-1 mice, *Int. J. Cancer,* 10, 489, 1972.
26. **Campbell, J., Richardson, A., and Schager, M.,** Insecticide residues in human diet, *Arch. Environ. Health,* 10, 831, 1965.
27. **Mrak, E.,** Report of the Secretary of Commission on Pesticides and their relationship with mental health, U.S. Department Health, Education and Welfare, Washington, D.C., 1969, 677.
28. **Davies, J., Edmundson, W., Edmundson, A., and Faffonell, A.,** The role of house dust in human DDT pollution, *Am. J. Publ. Health,* 2, 15, 1975.
29. **Smith, D. C., Leduc, R., and Tremblay, L.,** Pesticides residues in the total diet in Canada IV-1972 and 1974, *Pest. Series,* 6, 75, 1975.
30. **Gray, W. E., Marthre, D. E., and Rogers, S. J.,** Potential exposure of commercial seed-treating applicators to the pesticides carboxin-thiram and lindane, *Bull. Environ. Contam. Toxicol.,* 31, 2, 244, 1983.
31. Lindane, Data Sheets on Pesticides No. 12, World Health Organization/Food and Agriculture Organization unpublished document, 1975.
32. **Greve, P. A.,** Environmental and human exposure to hexachlorbenzene in the Netherlands, in *Hexachlorbenzene,* Proc. Int. Symp., Morris, C. R. and Carbaral, J. R. P., Eds., IARC SC. Publ. No. 77, IARC, Lyon, 1986.

33. **Brwon, E. A., Biddle, K., and Spaulding, J. E.,** Residue levels of hexachlorbenzene in meat and poultry in the food supploy of the USA, in *Hexachlorbenzene,* Proc. Int. Symp., Morris, C. L. and Carbaral, J. R. P., Eds., IARC Sc. Publ. No. 77, IARC, 1986.

34. **Uhnák, J., Venigerová, M., Madaričv, A., and Szokolay, A.,** Dynamics of hexachlorbenzene residues in the food chain, in Proc. Int. Symp., Morris, C. R. and Carbaral, J. R. P., IARC Sc. Publ. No. 77, IARC, Lyon, 1986.

35. **Nijhuis, H. and Heeschen, W.,** Hexachlorbenzene contamination of milk and human samples, in Proc. Int. Symp., Morris, C. R. and Carbaral, J. R. P., IARC Sc. Publ. No. 77, IARC, Lyon, 1986.

36. **Weisenberg, E.,** Hexachlorbenzene in human milk: a polyhalogenated risk, in Proc. Int. Symp., Morris, C. R. and Carbaral, J. R. P., IARC Sc. Publ. No. 77, IARC, Lyon, 1986.

37. Chlordane, EHC No. 34, World Health Organization, Geneva, 1986.

38. **Osetrov, V. I.,** Occupational hygiene in using heptachlor in agriculture, *Vrač. Delo.,* 3, 297, 1960.

39. **Kaloyanova, F.,** Organochlorine pesticides, in Toxicology of Pesticides, Interim Document No. 9, World Health Organization Regional Office for Europe, Copenhagen, 1982, 123.

40. **Fishbein, L.,** Chromatographic and biological aspects of DDT and its metabolites, *J. Chromat.,* 98, 177, 1974.

41. **Fisherova-Bergerova, V., Radomski, J. V., and Davies, J.,** Levels of hydrocarbon pesticides in human tissues, *Ind. Med. Surg.,* 36, 65, 1967.

42. **Dale, W., Coleland, M., and Hayes, W.,** Chlorinated pesticides in the body fat of people in India, *Bull. OMS,* 33, 471, 1964.

43. **Wassermann, M., Francone, M., Wassermann, D., Groner, Y., and Mariani, F.,** Accumulation des insecticides organochlorés dans les tissues adipeux de la population générale d'Argentine, *Sem. Med.,* avr., 459, 1969.

44. **Wassermann, M., Gon, M., Wassermann, D., and Zeller-Mayer, L.,** DDT and DDE in the body fat of people in Israel, *Arch. Environ. Health,* 11, 375, 1965.

45. **Bick, M.,** The effect of blood cholinesterase activity of chronic exposure to pesticides, *Med. J. Austr.,* 27, 1066, 1967.

46. **Wassermann, M., Curhov, D., Forte, P., and Groner, Y.,** Storage of organochlorine pesticides in the body fat of people in West Australia, *Int. J. Ind. Med. Surg.,* 37, 4, 295, 1968.

47. **Robinson, J., Richardson, A., Hunter, C., Crabtree, A., and Rees, H.,** Organochlorine insecticide content of human adipose tissue in Soud-Eastern England, *Br. J. Industr. Med.,* 22, 3, 220, 1965.

48. **Cassida, W., Fisher, A., Peden, I., and Parry Iones, A.,** Organochlorine pesticide residues in human fat from Somerset. *Monthly Bull. Min. Health and Publ. Health. Lab. Service,* 26(a), 2, 1967.

49. **Abbot, D., Goulding, R., and Tatton, J.,** Organochlorine pesticide residues in human fat in Great Britain, *Br. Med. J.,* 3, 146, 1968.

50. **Hunter, C. and Robinson, J.,** Pharmacodynamics of dieldrin (HEOD). Part I., *Arch. Environ. Health,* 15, 614, 1967.

51. **Fernandez, J., Astolfi, E., de Juarez, M., and Piacantino, H.,** Chlorinated pesticides found in the fat of children in the Argentine Republic, Amsterdam Conf., *Epidem. Toxicol. Pest.,* 23, 1971.

52. **Kaloyanova, F., Mihailova, Z., Georgiev, G., Benchev, I., Rizov, V., and Velichkova, V.,** Organochlorine pesticides in the fat of the general population of Bulgaria, Abstracts 17th Int. Congr. Occupational Health, Buenos Aires, 1972, 75.

53. **Rizov, N. A.,** Organochlorine pesticides in the fat and blood of the general population of Bulgaria and methods for their determination, Dissertation, Institute of Hygiene and Occupational Health, Sofia, 1977, (in Bulgarian).

54. **Wassermann, M.,** Present stage of the storage of organochlorine pesticides in the general population of South Africa, *S. A. Med. J.,* 44, 30, May, 646, 1970.

55. **Maier-Bode, H.,** DDT in Körperfett des Menschen, *Med. Exp.,* 1, 146, 1960.

56. **Engst, R., Knoll, R., and Nickel, B.,** Concentrations of chlorinated hydro-carbons, especially DDT and its metabolite DDE, in human fat, *Pharmazie,* 22, 654, 1967.

57. **Engst, R. and Knoll, R.,** Organochlorine residues in human milk, *Pharmacie,* 27, 52, 1972.

58. **Wasermann, M., Wassermann, D., and Cucos, S.,** Storage of polychlorinated bephenils in people of Israel, *ACRM,* 208, 1973.

59. **Ramachandran, M., Sharma, M., Sharma, S., Mathur, P., Aravindaksha, A., and Eduard, G. I.,** DDT and its metabolites in the body fat of Indians, *J. Com. Diseases,* 6, 256, 1974.

60. **Martines Leinares, V. and Wassermann, M.,** Almacenamiento de DDT en el paniculo adiposo de la populacion espanola, *Med. Segur. Trab.,* 15, 59, 29, 1967.

61. **Paccagnella, B., Prati, J., and Cavazzini, G.,** Chlorinated hydrocarbon insecticides in adipose tissue of persons of the province of Ferrada, *Nuovi Ann. Ig. Microbiol.,* 18, 17, 1967.

62. **Del Vecchio, V. and Leoni, V.,** Determination of chlorinated hydrocarbon insecticides in biological material. Part III. Chlorinated hydrocarbon in adipose tissue of some groups of the Italian population, *Nuovi Ann. Ig. Microbiol.,* 18, 104, 1967.

63. **Read, S. and McKinley, W.,** DDT and DDE content of human fat, *Arch. Environ. Health,* 3, 209, 161.

64. **Mastromatteo, E.,** Pesticides in man's health. The picture in Ontario, in Conf. Epid. Toxicology Pesticide, Amsterdam, 1971a, 25.

65. **Pomorska, K. and Szucki, B.,** Polskie badania nad zawartoecia DDT w tkance tleszcowe, *Medycyna Wiejska,* 2, 119, 1971.

66. **Juszkiewicz, T. and Stec, J.,** Posostolosci incektycydow polichlorowi v tkance fluszczowy rolnikow wojewodzine babielskiego, *Polski Tyg. Lek.,* 26, 13, 462, 1971.

67. **Vas'kovskaja, M. and Komarova, M.,** *Materials in hygiene and toxicology,* in Gorogskja naucnaja konferencia molodyh uconyh (Kiev), 1967, 67, (in Russian).

68. **Graceva, G.,** On the question of the storage of pesticides in human organism, *Gig. Primen. Toksikol. Pestits. Klinik. Otravlenii,* 1967, 67, (in Russian).

69. **Dale, W. and Quiby, G.,** Chlorinated insecticides in the body fat of people in the United States, *Science,* 142, 3591, 593, 1963.

70. **Hoffman, W., Fishbien, W., and Andelman, M.,** The pesticide content of human fat tissue, *Arch. Environ. Health,* 9, 3, 387, 1964.

71. **Hoffman, W., Adler, H., Fishbein, W., and Bauer, I.,** Relation of pesticide concentration in fat to pathological changes in tissues, *Arch. Environ. Health,* 15, 758, 1967.

72. **Durham, W.,** Pesticide exposure levels in men and animals, *Arch. Environ. Health,* 6, 10, 842, 1965.

73. **Morgan, D. and Roan, C.,** Chlorinated hydrocarbon pesticide residue in human tissues, *Arch. Environ. Health,* 20, 4, 452, 1970.

74. **Kuts, F., Vobs, A., Johnson, W., and Wiersma, G.,** Mirex residues in human adipose tissue, *Environ. Entomol.,* 3, 882, 1974.

75. **Deichmann, W.,** The chronic toxicity of organochlorine pesticides in man, *V. N. Intern. Med. Book Corp.,* 347, 1973.

76. **Wassermann, M., Trishnanandd, M., Tomatis, L., Day, N., Wassermann, D., Rungpitarasi, V., Chiamsakol, V., Djavaherian, M., and Cucos, S.,** Storage of organochlorine insecticides in adipose tissue of people in Thailand, *Southeast Asian J. Trop. Med. Publ. Health,* 3, 280, 1972.

77. **Wassermann, D., Tomatis, L., Wassermann, D., Day, N., and Djavaherian, M.,** Storage of organochlorine insecticides in adipose tissue of people in Uganda, *Bull. Environ. Contam. Toxicol.,* 12, 501, 1974.

78. **Denes, A.,** Lebensmittelchemische Probleme von Rückständen chlorierter Kohlenwasserstoffen, *Die Nahrung,* 6, 48, 1962.

79. **Hayes, W., Dale, W., and Berton, R.,** Storage of insecticides in French people, *Nature,* 21, 1189, 1963.

80. **Fournier, E.,** Phénomènes biochimiques observés au niveau du foie au cours des hépatites toxique, in Comptes rend. 7e réunion nation. des Centres de luttes contre les poisons, Masson, Paris, 1, 1966, 12.

81. **De Vlieger, M., Robinson, J., Baldwin, M., Crabtree, A., and Van Dijk,** The organochlorine pesticide content of human tissues, *Arch. Environ. Health,* 17, 759, 1968.

82. **Halacka, H., Hakl, J., and Vymetal, F.,** Effect of massive doses of DDT on human adipose tissues, *Cesk. Hyg.,* 10, 188, 1965.

83. **Rosival, L., Szokolay, A., Gorner, E., Madaric, A., and Uhnak, J.,** Exposure to pesticides and protection of health of the population, Whither Rural Med., Proc 4th Int. Congr. Rural Med., 1969, Usuda, Tokyo, 1970, 40.

84. **Adamovic, V., Hut, M., Sisindzić, M., and Djukić, V.,** Kumulacja organochlornich insecticida u nekim organisma i measnom tkivu stanovnika Srebije, *Hrana Ishrana,* 11, 1, 12, 1970.

85. **Kasai, A.,** Studies on the analysis of pesticide residues in the human body on foods. Whither Rural Medicine, Proc. 4th Int. Congr. Rural Med. Usuda, 1969, Tokyo, 1970, 86.

86. **Tatsumi, M., Sugaya, T., Sasaki, S., Suzuki, Y., and Satot,** Pesticide residues in human milk and fat tissues, in Abstr. 5th Congr. Int. Rural Med., (Varna), 1972, 122.

87. **Durham, W.,** The interaction of pesticides with other factors, *Residue Reviews,* 18, 21, 1967.

88. **Hoffman, W.,** Clinical evaluation of the effect of pesticides to man, *Ind. Med. Surg.,* 37, 289, 1968.

89. **Morgan, D. and Roan, C.,** Renal function in persons occupationally exposed to pesticides, *Environ. Health,* 19, 5, 633, 1969.

90. **Hayes, W., Durham, I., and Cueto, C.,** The effect of known repeated oral doses of chlorophenothane (DDT) in man, *JAMA,* 162, 1956, 890.

91. **Durham, W., Armstrong, L., and Quinby, G.,** DDT and DDE content of complete prepared meals, *Arch. Environ. Health,* 11, 5, 641, 1965.

92. **Hunter, C., Robinson, J., and Jager, K.,** Aldrin and dieldrin. The safety of present exposure of the general population of the UK and the USA, *Fd. Cosm. Toxicol.,* 5, 781, 1967.

93. **Hunter, C., Robinson, J., and Roberts, M.,** Pharmacodynamics of dieldrin (HEOD). Part II., *Arch. Environ. Health,* 13, 12, 1969.

94. **Robinson, J.,** The relationship between the dietary and body burden of HEOD (dieldrin), *Eur. J. Toxicol.,* 2, 210, 1969.

95. **Bertram, H. P., Kemper, F. H., and Müller, C.,** Hexachlorbenzene content in human whole blood and adipose tissue: experiences in environmental specimen banking, in *Hexachlorobenzene,* Proc. Int. Symp., Morris, C. R. and Carbaral, J. R. P., Eds., *IARC Sc. Publ.,* IARC, Lyon, 1986, No. 77.

96. **Robinson, P. E., Leczynski, B. A., Kurtz, F. W., and Remmers, J. C.,** An evaluation of hexachlorbenzene body-burden levels in the general population of the USA, in *Hexachlorobenzene,* Proc. Int. Symp., Morris, C. R. and Carbral, J. R. P., Eds., *IARC Sc. Publ.,* IARC, Lyon, 1986, No. 77.

97. **Radomsky, J., Deichmann, W., and Clizer, E.,** Pesticide concentrations in the liver, brain and adipose tissue of terminal hospital patients, *Fd. Cosmet. Toxicol.,* 6, 209, 1968.

98. **Casaretti, L., Freyer, G., Yauger, W., and Klemmer, H.,** Organochlorine pesticides residues in human tissue, *Hawaiian Environ. Health,* 17, 93, 306, 1968.

99. **Wyllie, J., Gabica, I., and Benson, W.,** Organochlorine pesticide residues in serum and biopsied lipid, *Pest. Monit. J.,* 6, Sept. 2, 84, 1972.

100. **Cueto, C. and Biros, F.,** Chlorinated insecticides and related material in human urine, *Toxicol. Appl. Pharmacol.,* 102, 261, 1967.

101. **Polishuk, Z., Wassermann, M., Wassermann, D., Groner, Y., Lazarovici, S., and Tomatis, L.,** Effects of pregnancy on storage of organochlorine insecticide, *Arch. Environ. Health,* 20, 215, 1970.

102. **Cureley, A. and Kimbrough, R.,** Chlorinated hydrocarbon insecticides in plasma and milk of pregnant and lactating women, *Arch. Environ. Health,* 18, 156, 1969.

103. **Astolfi, G., Alonso, A., and Mendizabol, A.,** Pesticides chlorés de l'accouchée et du cordon ombilical des nouveau-nés, *J. Europ. Toxicol.,* 7, Sept.-Dec. 5-6, 331, 1974.

104. **Trebicka-Kwiatkowska, B., Radomanski, T., and Stec, J.,** Stezenie DDT i jogo metabolitów w tkance tozyskowy i krwizyenej zylany kobirt z ciaza obumarta, *Polski Tugodnink Lekarski,* 30, 42, 1169, 1974.

105. **Dale, W., Curley, A., and Gueto, C.,** Hexane extractable chlorinated insecticides in human blood, *J. Sci.,* 5, 47, 1966.

106. **Nachman, G.,** A simplified method for analysis of DDT and DDE for epidemiological purposes, *Health Lab. Sci.,* 6, 148, 1969.

107. **Apple, J., Morgan, I., and Roan, C.,** Determination of serum DDT and DDE concentrations, *Bull. Environ. Contam. Toxicol.,* 5, 1, 16, 1970.

108. **Long, K.,** Pesticide residues in the blood of migrant field workers in relation to occupational exposure, in Whither Rural Med., Proc. 4th Int. Congr. Rural Med., Usuda, Tokyo, 1970, 71.

109. **Radomski, I., Astolfi, E., Deichmann, W., and Alberto, A.,** Blood levels of organochlorine pesticide in Argentine — occupationally and nonoccupationally exposed adults, children and newborn infants, *Tox. Appl. Pharmacol.,* 20, 186, 1971.

110. **Kasai, A.,** Organochlorine insecticides residues in human bodies versus their residues in food, in Abst. 5th Congr. Int. Rural Med., Varna, 1972, 119.

111. **Zavon, M.,** Blood cholinesterase levels in organic phosphate intoxication, *JAMA,* 192, 1, 137, 1965.

112. **Curley, A., Copeland, M., and Kimbrough, R.,** Chlorinated hydrocarbon insecticides in organs of stillborn and blood of new babies, *Arch. Environ. Health,* 19, 628, 1968.

113. **Komarova, L. and Vas'kovskaja, L.,** Biological significance of the excretions of DDT with the milk of lactating women, *Gig. Primen. Toksikol. Pestits. Klin. Otravlenii.,* 458, 1968, (in Russian).

114. **Komarova, L.,** Intruteral and postnatal storage of organochlorine pesticides in tissues and its significance in neonatal pathology, *Gig. Primen. Toksikol. Pestits. Klin. Otravlenii.,* 445, 1973, (in Russian).

115. **Egan, H., Goulding, R., Roburn, J., and Tahon, J.,** Organochlorine pesticide residues in human fat and human milk, *Brit. Med. J.,* 3453, 66, 1965.

116. **Heyndrickx, A. and Mayers, R.,** The excretion of chlorinated hydrocarbon insecticides in human mother milk, *J. Pharmacol. Belgique,* 24, 459, 1969.

117. **Juszkiewicz, T., Janstec, T., Radomanski, T., and Trebicka-Kwiatkowska, B.,** Pozostalosci insektycydow polichlorowych w swarze i mleku kobiet po porodzie, *Polska Lek.,* 27, 17, 616, 1972.

118. **Hornabrook, R., Dyment, P., Gomes, G., and Wieseman, I.,** DDT residues in human milk in New Guinea natives, *Med. J. Australia,* 1, 1297, 1972.

119. **Acker, L. and Schulte, E.,** Über das Vorkommen chlorierter Kohlenwasserstoffe in menschlichem Fettgewebe und Human Milch, *Dtsch. Lebensmittel-Rundschau,* 11, 385, 1970.

120. Heptachlor, EHC No. 38, World Health Organization, Geneva, 1984.

121. Chlordane, EHC No. 34, World Health Organization, Geneva, 1984.

122. **Slorach, S. A. and Vaz, R.,** Assessment of human exposure to selected organochlorine compounds through biological monitoring, GEMS Prepared for the United Nations Environment Programme and the World Health Organization by the Swedish National Food Administration, Upsala, 1983.

123. **Olszyna Marzys, A. E.,** Contaminants in human milk, *Act Pediatr. Scand.,* 67, 571, 1978.

124. **Lesby, L., Kowell, K., Houser, A., and Junker, G.,** Comparison of chlorinated hydrocarbon pesticides in maternal blood and placental tissues, *Environ. Res.,* 2, 247, 1969.

125. **Laws, E., Curley, A., and Biros, F.,** Men with intensive occupational exposure to DDT. A clinical and chemical study, *Arch. Environ. Health,* 15, 766, 1976.

126. **Edmundson, W., Davies, J., Cranner, M., and Nachman, G.,** Levels of DDT and DDE in blood and DDA in urine of pesticides formulators following a single intensive exposure, *Ind. Med. Surg.,* 38, 145, 1969.

127. **Edmundson, W., Davies, J., Nachman, G., and Roeth, R.,** P,P'-DDT and P,P'-DDE in blood samples of occupationally exposed workers, *Publ. Health Rep.,* 84, 53, 1969.

128. **Perron, R. and Barrentine, B.,** Human serum DDT concentrations related to environmental DDT exposure, *Arch. Environ. Health,* 20, 3, 368, 1970.

129. **Hunter, C. and Robinson, J.,** Aldrin, dieldrin in man, *Fd. Cosm. Toxicol.,* 6, 259, 1968.

130. **Yonchev, D. and Bardarov, V.,** DDT content in blood of pilots from agricultural aircraft, *Hygiena i Zdraveopazvane,* 23, 3, 220, 1980, (in Bulgarian).

131. **Griffith, J. and Duncan, R. C.,** Serum organochlorine residues in Florida citrus workers compared to the national health and nutrition survey examination sample, *Bull. Environ. Contam. Toxicol.,* 35, 3, 411, 1985.

132. **Jager, K.,** *Aldrin, dieldrin, endrin and telendrin. An epidemiological and toxicological study of long-term occupational exposure.* Elsevier, Amsterdam, 1970, 234.

133. **Robinson, J. and Roberts, M.,** Estimation of the exposure of the general population to dieldrin (HEOD), *Fd. Cosm. Toxicol.,* 501, 1969.

134. **Clifford, N. and Weil, J.,** Cortisol metabolism in persons occupationally exposed to DDT, *Arch. Environ. Health,* 24, 2, 145, 1972.

135. **Morgan, D. and Roan, C.,** Adrenocortical function in persons occupationally exposed to pesticides, *JOM,* 15, 1, 26, 1973.

136. **Bleherman, N.,** Excretion of steroid hormones in persons exposed to polychlorpinen, polychlorcamphen, hexachlorane, tetramethylthiuramdisulphide, *Gig. Primen. Toksikol. Pestits. Klin. Otravlenii,* 437, 1973, (in Russian).

137. **Hunter, C.,** Allowable human body concentrations of organochlorine pesticides, *Med. Lav.,* 577, 1968.

138. WHO, TRS N. 560, 1975.

139. **Eskenasy, J. J.,** Status epilepticus by dichlorodipenyltrichloroethane and hexachlorocyclohexane poisoning, *Rev. Roum. Neurol.,* 9, 435, 1972.

140. **Kluge, W. and Olbrich, H.,** Poisoning by a DDT-lindane combination, *Z. Hertzl. Fortbild,* 66, 980, 1972.

141. **Krasnyuk, E.,** Clinical forms of diencephalonpathology in subjects working with organochlorine compounds. I., *Gig. Primen. Toksikol. Pestits. Klin. Otravlenii,* 336, 1973, (in Russian).

142. **Mayersdorf, A. and Israeli, A.,** Toxic effects of chlorinated hydrocarbon insecticides on the human electroencephalogramme, *Arch. Environ. Health,* 28, 159, 1974.

143. **McIntire, M., Angle, C., and Maragos, S.,** Insecticide poisoning in childhood: follow up evaluation, *J. Pediat.,* 67, 647, 1965.

144. **Jenkins, S. and Toole, J.,** Polyneuropathy following exposure to insecticides, *Arch. Int. Med.,* 5, 113, 691, 1964.

145. **Chakravarti, H.,** Polyneuritis due to DDT, *J. Ind. Med. Assoc.,* 45, 598, 1965.

146. **Schuttmann, W.,** Polyneuritis nach beruflichen Kontakt mit DDT, *Z. Ges. Hyg.,* 12, 5, 307, 1966.

147. **Onifer, T. and Whisnant, J.,** Cerebral ataxia and neuritis after exposure to DDT and lindane, Proc. Staff Meet. Mayo Clinic, 30, 4, 1957.

148. **Model, A. and Zaritskaja, L.,** On the pathology of the vegetative part of nervous system by chronic intoxications, *Gig. Tr. Prof. Zabol.,* 8, 17, 1967, (in Russian).

149. **Bogusz, M.,** Studies on the activity of certain enzymatic systems in agriculture workers exposed to the action of OP insecticides, *Pol. Lek. Wiadonsci. Lek.,* 23, 786, 1968.

150. Committee on Pesticides, Pharmacologic and Toxicologic Aspects of DDT (chlorophenothane U.S.P.), *J. Am. Med. Assoc.,* 145, 728, 1951.

151. **Salihodzaev, S. S. and Ferstat, V. N.,** Condition of the olfactory analyzer under the action of organochlorine and organophosphorus pesticides, *Gig. Sanit.,* 37, 95, 1972, (in Russian).

152. **Loganovskij, N.,** The effect of organochlorine toxic chemicals on the functional state of kidneys, *Gig. Primen. Toksikol. Pestits. Klin. Otravlenii,* 528, 1968, (in Russian).

153. **Krasnyuk, E.,** ECG changes in subjects working with OC insecticides, *Sov. Med.,* 9, 28, 134, 1964, (in Russian).

154. **Laws, E. R., Jr., Curley, A., and Biros, F. J.,** Men with intensive occupational exposure to DDT. A clinical and chemical study, *Arch. Environ. Health,* 15, 766, 1967.

155. **Krasnyuk, E.,** Functional state of cardiovascular system in subjects working with organochlorines, *Gig. Primen. Toksikol. Pestits. Klin. Otravlenii,* 486, 1968, (in Russian).

156. **Boyko, V.,** State of the olphactory analyzer in subjects exposed to OC compounds, *Gig. Primen. Toksikol. Pestits. Klin. Otravlenii,* 538, 1968, (in Russian).

157. **Bezuglyj, V. and Buslenko, A.,** Heart rhythm disturbances with toxicological etiology, *Gig. Primen. Toksikol. Pestits. Klin. Otravlenii,* 406, 1973, (in Russian).

158. **Friberg, L. and Martensson, I.,** Case of panmyelophtisis after exposure to chlorphenotane and benzene hexachloride, *Arch. Ind. Hyd. Med.,* 8, 166, 1953.

159. **Mandeloff, A. and Smith, D.,** Insecticides, bone marrow failure, gastrointestinal bleeding and incontrollable infections, *Amer. J. Med.,* 19, 2, 275, 1955.

160. **Albahary, C., Dubrissay, J., and Querin,** Pancitopenie rebelle au lindane (isomère gama de IHCH), *Arch. Mal. Prof.,* 18, 6, 683, 1957.

161. **Kalyaganov, P. and Lashtenko, N.,** Some questions of the clinic of chronic intoxications with OC pesticides, *Gig. Primen. Toksikol. Pestits. Klin. Otravlenii,* 477, 1968, (in Russian).

162. **Komarova, L. and Batueva, L.,** Some data on the relation of blood system disease to the effect of OC pesticides, *Gig. Primen. Toksikol. Pestits. Klin. Otravlenii,* 389, 1973, (in Russian).

163. **Hrycek, A., Kalina, Z., and Oweczarzy, I.,** Cytochemical investigation of selected dehydrogenases in peripheral blood leukocytes in workers exposed to polychlorinated pesticides, *Med. Pr.,* 35, 3, 185, 1984, (in Polish).

164. **Krasnyuk, E.,** Combined effect of different insecticides on digestive organs, *Gig. Primen. Toksikol. Pestits. Klin. Otravlenii,* 532, 1965, (in Russian).

165. **Paramonchik, V.,** The state of liver excretory function in workers exposed to OC pesticides, *Gig. Primen. Toksikol. Pestits. Klin. Otravlenii,* 97, 1966, (in Russian).

166. **Paramonchik, V. and Platonova, B.,** On the functional state of liver in subjects exposed to organochlorine toxic chemicals, *Gig. Truda Prof. Zabol.,* 12, 3, 27, 1968, (in Russian).

167. **Platonova, V.,** Data on the secretory activity of stomach in workers exposed to organochlorine toxic chemicals, *Gig. Primen. Toksikol. Pestits. Klin. Otravlenii,* 515, 1968, (in Russian).

168. **Paramonchik, V.,** Effect of organochlorine toxic chemicals on some functions of liver, *Gig. Primen. Toksikol. Pestits. Klin. Otravlenii,* 412, 1968, (in Russian).

169. **Karimov, M.,** Free aminoacid content in serum of subjects with occupational contact with pesticides, *Gig. Primen. Toksikol. Pestits. Klin. Otravlenii,* 551, 1968, (in Russian).

170. **Morgan, D. and Roan, C.,** Liver function in workers having high tissue store of chlorinated hydrocarbon pesticides, *Arch. Environ. Health,* 29, 14, 1975.

171. **Kolmodin, B., Azarnoff, D. L., and Sjoquist, F.,** Effect of environmental factors on drug metabolism: decreased plasma half-life of antipyrin of workers exposed to chlorinated hydrocarbon insecticides, *Clin. Pharmacol. Ther.,* 10, 638, 1969, (in Russian).

172. **Poland, A., Smith, D., Kuntzman, R., Jacobson, M., and Conney, A. H.,** Effect of intensive occupational exposure to DDT on penylbutazone and cortisol metabolism in human subjects, *Clin. Pharmacol. Ther.,* 11, 724, 1970.

173. **Krasnyuk, E. P. and Platonova, V. I.,** Disturbances of stomach functions by long-term exposure to organochlorine toxic chemicals, Report of the Academy of Sciences of Uzbekistan, SSR, No. 9, 99, 1985.

174. **Southern, A. L., Tochimoto, S., Isurugi, K., Corcher, G. G., Krikun, E., and Stypulkowski, W.,** The effect of 2-2-bis-(2-chlorophenyl-4-chlorophenyl)-1,1,1-dichloroethane(o,p′-DDD) on the metabolism of infused cortisol-7-H, *Steroids,* 7, 11, 1966.

175. **Hellman, L., Bradlow, H. L. and Zymoff, B.,** Decreased conversion of androgens to normal 17-ketosteroid metabolites as a result of treatment with o,p′-DDD, *J. Clin. Endocrin. Metabol.,* 36, 801, 1973.

176. **Simic, B. and Kniewald, J.,** Effects of pesticides on the reproductive system, *Acta Pharm. Yugoslavia,* 30, 59, 1980.

177. **O'Leary, J. A., Davies, J. E., and Feldman, M.,** Spontaneous abortion and human pesticide residues of DDT and DDE, *Am. J. Obstet. Gynecol.,* 108, 1291, 1970.

178. **Wassermann, M., Ron, M., Bercowici, B., Wassermann, D., Cucos, S., and Pines, A.,** Premature delivery and organochlorine compounds: polychlorinated biphenyls and some organochlorine insecticides, *Environ. Res.,* 28, 1, 106, 1982.

179. **Pushkar, N. and Bryazgunov, V.,** Cases of intoxicaiton by products, contaminated with DDT, *Gig. Primen. Toksikol. Pestits. Klin. Otravlenii,* 340, 1966, (in Russian).

180. **Conley, E.,** Morbidity and mortality from economic poisoning in the United States, *Arch. Ind. Health,* 18, 2, 126, 1958.

181. **Kolarski, V.,** On the acute peroral intoxication with DDT, *Savremenna Medicina,* 12, 13, 45, 1962, (in Bulgarian).

182. DDT - Data sheet on Pesticides No. 21, World Health Organization/Food and Agriculture Organization, unpublished document, 1976.

183. **Model', A.,** Specificity of neurological symptomatics of the chronic intoxications by DDT, *Sov. Medicina,* 1, 110, 1968, (in Russian).

184. **Model', A.,** Some data on neurological symptomatics of chronic intoxications with toxic chemicals, *Gig. Primen. Toksikol. Pestits. Klin. Otravlenii,* 538, 1965, (in Russian).

185. **Hermann, B.,** Intoxication au DDT chez l'homme; un cas grave de poluneurite causé par le DDT, *Acta Med. Budapest,* 11, 209, 1958.

186. **Michon, P., Largan, A., Huriet, C., and Caucher, P.,** Polyneurite des membres intérieurs consécutive à l'absorption accidentelle d'une poudre de DDT, *Rev. Neurol.,* 106, 3, 325, 1962.

187. **Alexieva, T., Vassilev, G., and Spassovski, M.,** A study of the chronic effect of organochlorine pesticides — DDT and hexachlorane — upon humans, *Trudove na NIOTPZ,* 4, 115, 1959, (in Bulgarian).

188. **Misra, U. K., Nag, D., and Krishna Murti, C. R.,** A study of cognitive functions in DDT sprayers, *Ind. Health,* 22, 3, 199, 1984.

189. **Schuttmann, W.,** Chronic liver diseases after occupational exposure to dichlordiphenyl trichlorethane (DDT) and hexachlorocyclohexane (HCH), *Int. Arch. Gewerbepath. Gewerbehyd.,* 24, 193, 1968.

190. **Davidson, C.,** Quelque indications sur l'effet du chlorphénothane (DDT) sur le foie, *Rev. Int. Hepath.,* 2, 1968.

191. **Carlson, L. and Kolmodin-Hedman, B.,** Hyper-α-lipoproteinemia in men exposed to chlorinated hydrocarbon pesticides, *Acta Med. Scand.,* 192, 29, 1972.

192. **Wright, C., Donn, C., and Haymie, H.,** Agranulocytosis occuring after exposure to DDT, pyrethrum aerosol bomb, *Am. J. Med.,* 1, 562, 1946.

193. **Jonczyk, H., Rudowski, W., Traczyk, Z., and Klawe, Z.,** DDT and its metabolites and gamma-HCH in blood, bone marrow and fatty tissue in certain hematological syndromes, *Pol. Tyg. Lek,* 29, 1573, 1974, (in Polish).

194. **Karpinski, F.,** Purpura following exposure to DDT, *J. Pediatr.,* 31, 375, 1950.

195. **Heyndrickx, A. and Mayers, R.,** Two accidental cases of p,p′-DDT (dichlorodiphenyl trichlorpenthane) poisoning in children, *J. Pharmacie Belgique,* 24, 453, 1969.

196. **Krasnyuk, E., Loganovskij, N., Makovskaja, E., and Rappoport, M.,** Functional morphological changes in kidneys under exposure of the organism to DDT, *Sovetskaya Medicina,* 31, 6, 38, 1968, (in Russian).

197. **Linon, M.,** A propos de quelques accidets cutanés par le DDT en solution dans le pétrole, *J. Méd. Bordeaux,* 132, 6, 623, 1955.

198. **Bledsoe, T., Roland, D. P., Hey, R. L., and Liddle, G. W.,** An effect of o,p′-DDD on extra-adrenal metabolism of cortisol in man, *J. Clin. Endocrinol. Metab.,* 24, 1303, 1964.

199. **Rabello, M. N., Bacak, W., Almeida, W. F., Pigati, P., Ungaro, M. T., Murata, T., and Pereira, C. A. B.,** Cytogenetic study on individuals occupationally exposed to DDT, *Mutat. Res.,* 28, 449, 1975.

200. **Burkatskaya, E.,** Abstracts of papers of the Scientific Session on the problems of medical examination of workers from the chemcial industry (Gorkij), 1959, 88, (in Russian).

201. **Kalyaganov, P.,** On the clinic of chronic intoxications with technical hexachlorane, *Gig. Primen. Toksikol. Pestits. Klin. Otravlenii,* 546, 1965, (in Russian).

202. **Odintsova, I.,** Peculiarities of the disturbance of the functional state of pancreas at pathology of the hepatobiliary system due to long-term exposure to hexachlorane, *Gig. Primen. Toksikol. Pestits. Klin. Otravlenii,* 1973, (in Russian).

203. **Gorskaya, N., Odintsova, I., Soboleva, L., and Kazakevich, R.,** On the functional state of the cardiovascular system under long-term exposure to hexachlorane, *Gig. Primen. Toksikol. Pestits. Klin. Otravlenii,* 409, 1973, (in Russian).

205. **Wakatsuki, T.,** The actual state of pesticide poisoning of farmers in Japan, in 3rd Congr. Int. Med. Rural, 10, 1966, Bratislava, II, 79.

206. **Wassermann, M., Michail, G., Vancea, G., Mandrié, G., Hisseu, S., Roileanu, Y., Sava, V., Josubas, S., and Nestor, L.,** Recherches sur les conditions de milieu et sur la pathologie professionelle des désanophélisateures. L'intoxication chronique par l'hexachlorocyclohexane, *Arch. Mal. Prof.,* 23, 1962, 1, 2, 18.

207. **Marchand, M., Durbrull, P., and Goudenand, M.,** Agranulocytose chez un sujet soumis à des vapeurs d'hexachlorocyclohexane, *Arch. Mal. Prof.,* 17, 256, 1956.

208. **McLean, J.,** Aplastic anaemia associated with insecticides, *Med. J. Aust.,* 1, 53, 996, 1966.

209. **Woodliff, H., Connor, P., and Scopa, J.,** Aplastic anaemia associated with insecticides, *Med. J. Aust.,* 1, 53, 15, 628, 1966.

210. **Loge, J.,** Aplastic anaemia following exposure to benzene hexachloride (lindane), *J. Med. Assoc.,* 193, 2, 110, 1965.

211. **Tolot, F., Lenglet, J., Frost, G., and Bertholon, J.,** Polyneurite due au lindane, *Arch. Mal. Prof.,* 30, 443, 1969.

212. **Dési, I.,** Neurotoxicological effect of small quantities of lindane. Animal Studies, *Int. Arch. Arbeitsmed.,* 33, 153, 1974.

213. **Danopoulos, E., Melissinos, K., and Katsas, G.,** Serious poisonings by hexachlorocyclohexane. Clinical and laboratory observations on five cases, *Arch. Ind. Hyg. Occ. Med.,* 6, 6, 582, 1953.

214. **Starr, G. and Clifford, J.,** Acute lindane intoxication, *Arch. Environ. Health,* 25, 374, 1972.

215. **Bambov, C., Chomakov, M., and Dimitrova, N.,** Group intoxication with lindane, *Savr. Med.,* 17, 6, 447, 1966, (in Bulgarian).

216. **Osuntokun, B.,** "Gammelin 20" poisoning: a report of two cases, *W. Afr. Med. J.,* 13, 5, 207, 1964.

217. **Kay, R., Kuder, J., Sussleu, W., and Lewis, R.,** Fatal poisoning from ingestion of benzene hexachloride, *Ghana Med.,* 3, 2, 72, 1964.

218. **Kashyap, S. K., Gupta, S. K., Karnik, A. B., Perkin, J. R., and Chatterjee, S. K.,** Scope and need of toxicological evaluation of pesticides under field conditions — medical surveillance of marlaria spraymen exposed to HCH (hexachlorocyclohexane) in India, in *Field Worker Exposure During Pesticide Application,* Tordoir, W. F. and van Heemstra-Lequin, E. A. H., Eds., Elsevier, Amsterdam, 1980, 53.

219. **Patel, T. and Rao, V.,** Dieldrin poisoning in man. A report of 20 cases observed in Bombay State, *Br. Med. J.,* 5076, 921, 1958.

220. **Simanski, H.,** A case of fatal occupational dieldrin intoxications, *Ind. Med. Surg.,* 37, 7, 551, 1968.
221. **Hoogendam, J., Versteeg, J., and de Vlieger, M.,** Electroencephalograms in insecticide toxicity, *Arch. Environ. Health,* 4, 1, 86, 1962.
222. **Hoogendam, J., Versteeg, J., and de Vlieger, M.,** Nine years toxicity control in insecticide plants, *Arch. Environ. Health,* 10, 3, 441, 1965.
223. **Jacobs, P. and Lurie, J.,** Acute toxicity of the chlorinated hydrocarbon insecticides, *S. Afr. Med. J.,* 41, 1147, 1967.
224. **Spynu, E.,** On the assessment of the physiological changes on early stages of intoxications of dienic synthesis, *Gig. Truda Prof. Zabol.,* 10, 26, 1964, (in Russian).
225. **Pick, J., Joshua, H., Leffkowitz, M., and Gutman, A.,** Aplastic anemia following exposure to aldrin, *Harefuah,* 68, 5, 164, 1965.
226. **Kazantzis, G., McLaughlin, F., Prior, M.,** Poisoning in industrial workers by the insecticide aldrin, *Brit. J. Ind. Med.,* 21, 46, 1964.
227. **Coble, Y., Hildebrand, F., Davies, J., Raasch, F., and Curley, A.,** Acute endrin poisoning, *J. Am. Med. Assoc.,* 202, 439, 1967.
228. **Preda, J., Moraru, J., Manolescu, A., and Radovici, J.,** Aspects morphologiques de la intoxication aiguë par l'aldrine, *Ann. Méd. Lég.,* 43, 5, 483, 1963.
229. **Fokina, K.,** On the neurological characteristics of acute poisoning by polychlorocamphene, *Gig. Primen. Toksikol. Pestits. Klin. Otravlenii,* 382, 1973, (in Russian).
230. **Warraki, S.,** Respiratory hazards of chlorinated camphene, *Arch. Environ. Health,* 7, 2, 293, 1963.
231. Heptachlor - Data Sheet No. 19, 1975, World Health Organization/Food and Agriculture Organization, unpublished document, 1975.
232. **Garreston, L. and Curley, A.,** Dieldrin studies in a poisoned child, *Arch. Environ. Health,* 19, 814, 1969.
233. **Raalte Van, H.,** Aspects of pesticides toxicity, in papers presented at the Conf. Occup. Health, Caracas, Venezuela, 1965, 34.
234. **Aldrich, F. and Holmes, J.,** Acute chlordane intoxication in child, *Arch. Environ. Health,* 19, 129, 1969.
235. **Curley, A. and Gerrettson, L.,** Acute chlordane poisoning, *Arch. Environ. Health,* 18, 211, 1969.
236. **Stranger, J. and Kerridge, G.,** Multiple fractures of the dorsal part of the spine following chlordane poisoning, *Med. J. Aust.,* 1, 267, 1968.
237. **Fishbein, W. I., White, J. V., and Isaacs, H. J.,** Survey of workers exposed to chlordane, *Ind. Med. Surg.,* 33, 726, 1964.
238. **Soboleva, L.,** On the effect of polychlorpinen upon the cardiovascular system of man, *Gig. Primen. Toksikol. Pestits. Klin. Otravlenii,* 415, 1973, (in Russian).
239. **Homenko, N.,** Functional state of neuromuscular apparatus in people contacting with polychlorpinen, *Gig. Primen. Toksikol. Pestits. Klin. Otravlenii,* 386, 1973, (in Russian).
240. **Bezuglij, V., Kaskevich, L., Il'ina, V., Sivitskaja, I., and Bleherman, N.,** To the clinical characteristics of people exposed to polychlorpinen under conditions of its application, *Gig. Primen. Toksikol. Pestits. Klin. Otravlenii,* 412, 1973, (in Russian).
241. **Kulagin, O.,** The state of the protein synthetic function of liver in subjects with single exposure to polychlorpinen, *Gig. Primen. Toksikol. Pestits. Klin. Otravlenii,* 426, 1973, (in Russian).
242. **Ovsyannikova, L.,** Free aminoacids in peripheral blood and urine at polychlorpinen exposure, *Gig. Primen. Toksikol. Pestits. Klin. Otravlenii,* 402, 1973, (in Russian).
243. **Il'ina, V.,** The state of the menstrual function in subjects exposed to chlorpinen, *Gig. Primen. Toksikol. Pestits. Klin. Otravlenii,* 447, 1973, (in Russian).
244. **Shapovalov, Y.,** Experience with treatment of intoxication with methylbromide, *Gig. Primen. Toksikol. Pestits. Klin. Otravlenii,* 545, 1973, (in Russian).
245. **Adams, R., Zimmerman, M., Barlett, J., and Preston, J.,** 1-chloro-2,4-dinitrobenzene as an algidide. Report of four cases of contact dermatitis, *Arch. Dermatol.,* 103, 2, 191, 1971.
246. **Schmid, R.,** Cutaneous porphyries in Turkey, *N. Engl. J. Med.,* 263, 397, 1960.
247. **Stonard, M. and Nenov, P.,** Effect of hexachlorobenzene on hepatic microsomal enzymes in the rat, *Biochem. Pharmacol.,* 23, 2175, 1974.
248. **Gocmen, A., Peters, H. A., Cripps, D. J., Morris, C. R., and Dogramaci, I.,** Porphyria turcica: hexachlorobenzene-induced porphyria, in *Hexachlorobenzene,* Proc. Int. Symp., Morris, C. R. and Carbral, J. R. P., Eds., *IARC Sc. Publ.* No. 77, IARC, Lyon, 1986.
249. **Peters, H. A., Gocmen, A., Cripps, D. J., Morris, C. R., and Bryan, G. T.,** Porphyria turcica: hexachlorobenzene-induced porphyria. Neurological manifestations and therapeutic trials of ethylenediaminetetracetic acid in the acute syndrome, in *Hexachlorobenzene,* Proc. Int. Symp., Morris, C. R. and Carbral, J. R. P., Eds., *IARC Sc. Publ.* No. 77, IARC, Lyon, 1986.
250. **Peters, H. A., Cripps, D. J., Gocmen, A., Ertürk, E., Bryan, G. T., and Morris, C. R.,** Neurotoxicity of hexachlorobenzene-induced porphyria turcica, in *Hexachlorobenzene,* Proc. Int. Symp., Morris, C. R. and Carbral, J. R. P., Eds., *IARC Sc. Publ.* No. 77, IARC, Lyon, 1986.

251. **Peters, H. A., Cripps, D. J., Lambrecht, R. W., Ertürk, E., Morris, C. R., and Bryan, G. T.,** History and geography of hexachlorobenzene poisoning in Southeastern Turkey, in *Hexachlorobenzene,* Proc. Int. Symp., Morris, C. R. and Carbral, J. R. P., Eds., *IARC Sc. Publ.* No. 77, IARC, Lyon, 1986.
252. **Richter, E.,** Stimulation of faecal excretion of hexachlorobenzene, review, in *Hexachlorobenzene,* Proc. Int. Symp., Morris, C. R. and Carbral, J. R. P., Eds., *IARC Sc. Publ.* No. 77, IARC, Lyon, 1986.
253. Hexachlorobenzene. Data Sheets on Pesticides No. 26, 1977, World Health Organization/Food and Agriculture Organization, unpublished document, 1977.
254. **Kaloyanova-Simeonova, F.,** *Pesticides Toxic Action and Prophylaxis,* Bulgarian Academy of Science, Sofia, 1977.
255. WHO Technical Report Series 677. Recommended health-based limits in occupational exposure to pesticides, World Health Organization, Geneva, 1982.
256. NIOSH Pocket Guide to Chemical Hazards, Sept. 1985, United States Department of HHS, NIOSH.

Chapter 5

SYNTHETIC PYRETHROIDS

I. INTRODUCTION

Natural pyrethrins are constituents of flower extracts, *Pyrethrum cinerariae folium* and other species. The principal active constituents are pyrethrin I and II, cinerin I and II, and jasmolin I and II. They are esters of three alcohols — pyrethrolone, cinerolone, and jasmolone with chrysanthemic and pyrethric acids. Since 1973 similar compounds have been synthesized under the general name pyrethroids. They are similar to pyrethrin I from the natural esters. More than 1000 pyrethroids have been synthesized.

Synthetic pyrethroids (SP) possess greater insecticidal activity and lower mammalian toxicity than the organochlorine, organophosphate and carbamate insecticides. Due to their excellent cost/benefit ratio for pest control, SPs attained approximately 30% of the commercial market.

II. PROPERTIES

Synthetic pyrethroids are esters of specific acids: chrysanthemic acid, halogen-substituted chrysanthemic acid, 2-(4-chlorophenyl)-3-methylbutyric acid, and alcohols (e.g. allethrolone, 3-phenoxybenzyl alcohol). The most important compounds are given in Tables 1 and 2. Many of them are mixtures of different isomers.

The active substances are viscous liquids, they are highly lipophylic with high boiling points (120 to 160°C) and very low vapor pressure. They are not soluble in water, but they are soluble in organic solvents.

The alpha-cyano derivatives (II group) with CN group in the *S*-configuration of the 3-phenoxybenzyl alcohol (deltamethrin, cypermethrin, cyhalothrin, fenvalerate, fenpropathrin) have about three- to sixfold higher potency for insects than the noncyano derivatives (I group) permethrin, alphamethrin, bioallethrin.

Synthetic pyrethroids are formulated as emulsive concentrates (EC), flowable concentrates (FC), ULV, and WP formulations.

III. USES

Synthetic pyrethroids are among the most active insecticides. At concentrations of 0.02 to 0.05% (5 to 200 g/ha) they have the same effect as organophosphorous compounds at 0.5%. Some of them (fenvalerate, fenpropathrin, and cyhalothrin) are acaricides.

They are used in fields, orchards, and greenhouses, against domestic insects and animal pests, in stored products for seed dressing, etc. Pyrethroids are active against many pests resistent to other insecticides.

The harvesting period for synthetic pyrethroids varies from 0 to 14 d. Synthetic pyrethroids are easily decomposed in the environment (for 1 to 3 weeks), especially by ultraviolet radiation and higher temperature. Roberts et al. found that in sandy clay and sandy loam soils, cypermethrin degraded for 4 and 2 weeks, respectively.[1]

IV. METABOLISM

Synthetic pyrethroids are generally metabolized in mammals through ester hydrolysis, oxidation, and conjugation. There is no tendency to accumulate in tissues as an intact synthetic

TABLE 1
Synthetic Pyrethroids

Structural formula	Names

I. Group

CH₃ CH₃
Cl₂C=CH—CH—CH—COO—CH₂—⬡—O—⬡

Permethrin, Ambush, Ambushfog, Biopermethrin, Coopex, Dragnet, Corsair, Eksmin, Kafil, Outflank, Picket, Peregin, Pethrin, Pramex, Pounce, Stockard, Stomexin, Talcord, Tornado, Permasect

(CH₃)₂C=CH—CH—CH—COO—CH₂—furan—CH₂—⬡

Chismethrin, Resmethrin, Bioresmethin, Benzofuroline, Synthrin, Benzyfuroline, Chryson, Biobenzyfuroline, Pynosect

(CH₃)₂C=CH—CH—CH—COO—CH—...—CH₂—CH=CH₂

Allethrine, Isathrin, Pallethrin, Pynamin, Allycinerin, Exbrothrin, Pyresin, Necardboxyl acid, Esbiol

(CH₃)₂C=CH—CH—CH—COO—CH₂—⬡—O—⬡

d-Phenothrin, Symethrin, Pesgard

(CH₃)₂C=CH—CH—CH—COO—CH₂—N...

Tetramethrin, Phtalthrin Neopynamin

pyrethroid molecule. Biotransformation mainly takes place in liver microsomes, and the metabolites are eliminated in 2 to 4 d via urine, feces, and expired air. In humans 6 to 22 metabolites are isolated from 1 active compound, according to Eadsforth and Baldwin.[2]

Experimental studies have shown a covalent binding of pyrethroids and their metabolites to hepatic proteins with tropism for the smooth endoplastic reticulum. The latter explains the increased liver weight, elevated activity of hepatic aminopyrine demethylase, and the proliferation of smooth endoplasmic reticulum in the liver. The increase of microsomal cytochrome P-450 and NADTH cytochrome C reductase is small compared to phenobarbital.

V. TOXICITY: MECHANISM OF ACTION

Acute toxicity parameters vary in connection with the chemical structure and isomerism of the individual compounds and the vehicle used, as well as the species, sex, age, and diet of the animals used. The values reported sometimes differed markedly (more than 10 times) (Table 3).

Synthetic pyrethroids are nervous poisons. They interfere with the ionic permeability of the cell membranes involved with generation and conductance of nerve impulses. They act upon the axons of the peripheral and central nervous system; they displace the specific binding of kainic

TABLE 2
Synthetic Pyrethroids

Structural formula	Names

II. Group with CN Group in the Molecule

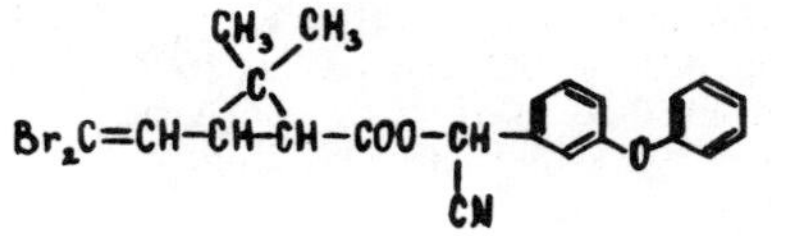

Deltamethrin,
Decamethrin,
Decis, Kothrin, Decasol

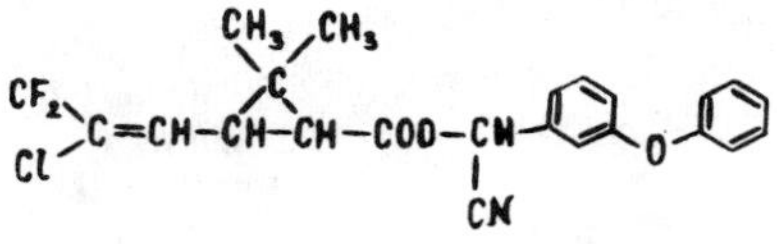

Cyhalothrin, Grenade, Karate

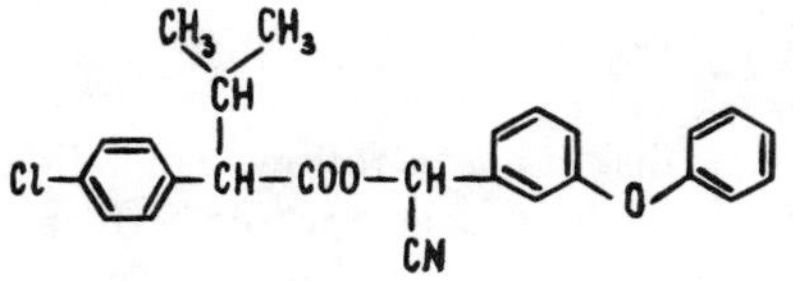

Fenvalerate,
Sumicidin
Belmark, Pydrin

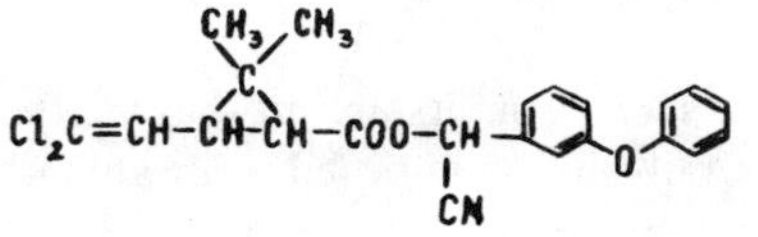

Cypermethrin,
Ripcord, Fastak,
Cymbush, Bastox,
Alphamethrin, Supersect,
Baricard, Aremo,
Cyperkill

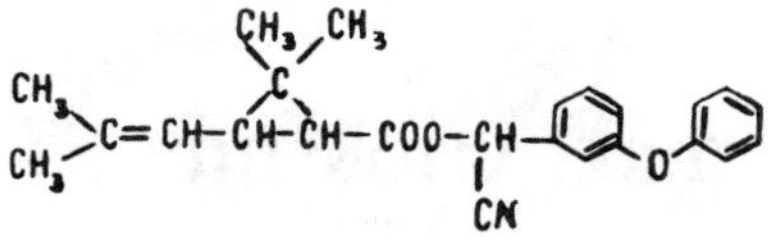

Fenpropathrin,
Danitol
Meothrin

acid in mammals by interacting with sodium channels gating mechanism. Synthetic pyrethroids induce the sodium channels to close slower than normal. The slow influx of sodium ions at the termination of depolarization is referred as a sodium tail current. Noncyanic pyrethroids induce tail currents with time decay on the order of milliseconds, alpha-cyano pyrethroids — hundreds of milliseconds to more than one second.[3]

The mechanism of sensory irritation (paraesthesia) due to percutaneous contact with synthetic pyrethroids is not thoroughly clear.

Based on the clinical symptoms of poisoned rats, two distinct signs of intoxication have been designated (T- and CS-syndrome) by Verschoyle and Aldridge.[4]

The T-syndrome of poisoning normally has a rapid onset with aggressive sparring and increasing sensitivity to external stimuli. This is followed by a fine tremor that gradually becomes more severe and generalized until the animals become prostrate with a coarse whole body tremor. Body temperature rises. Just prior to death (usually 1 to 2 d after beginning the intoxication), a rapid onset of rigor may be observed. It has been suggested that the T-syndrome is related mainly to an action on the peripheral nervous system, and the enhanced sensitivity is related to the lowering of the threshold of sensory receptors, resulting in repetitive activity in sense organs and sensory nerve fibers. The central nervous system is also involved.

TABLE 3
Toxicity of Pyrethroids[5]

Pyrethroids	Oral LD50 mg/kg/rats	Dermal LD$_{50}$ mg/kg	Inhalatory LC50 mg/m^3	NEL (diet)
Permethrin	430–7000	>700, >5000		
Resmethrin	8000	>10,000		3000 mg/kg rats 90 d
Bioresmethrin	8000	>10,000		1200 mg/kg rats 7500 mg/kg dog
Allethrin	685–1100	>2500		
Bioallethrin	405–1545	>2500	1500 mg/m^3 mice 1650 mg/m^3 rats	1000 mg/kg 6 mo rats
Tetramethrin	>5000	>5000		1500 mg/kg rats
d-Phenothrin	>5000	>10,000	>3760	2500 mg/kg, 6 mo
Deltamethrin	135–5000	>2000	600, 785	12 mg/kg mice 2.1 mg/kg rats
Fenvalerate	451–3200	Rat >5000; rabbit 1000–3200	>101 mg/m^3/3h	250 mg/kg rats
Cypermethrin	79–5000			
Fenpropathrin	67–71	870–1000		

The CS-syndrome is characterized by initial behavioral disturbances, followed by salivation within 2 to 5 min, myosis, bradycardia, tremors in the extremities, coarse whole body tremor, increased startle response, and abnormal locomotion involving the hind limbs. The tremor progresses into a choreoathetosis and clinic seizures. The CS-syndrome has been shown by Staatz et al to give rise to EEG changes indicative of a central action.[5] Finally, clonic convulsion of central genesis may appear.

Surviving animals normally recover in several hours. *cis*-Isomers are more toxic than the corresponding *trans*-isomers. The CS-syndrome has only been observed with alpha-cyano derivatives as deltamethrin, cyhalothrin, cypermethrin, fenvalerate, etc. The threshold dose for salivation, tremor, and ataxia has been determined by Brodie et al. to be 2.5 mg/kg deltamethrin i.p. and for spontaneous choroatheosis — 5 mg/kg.[6] Elevated cereballar cyclic GMP levels were found.

VI. EXPOSURE: DOSE-EFFECT RELATIONSHIP

People can be exposed to SPs during application and production, as well as by food.

Exposure to synthetic pyrethroids was studied in 17 persons during conifer seedling treatment and planting by Kolmodin-Hedman et al.[17] Fenvalerate, permethrin wettable powder, and permethrin emulsion concentrate were diluted with water to 1 to 2% active ingredient. A man dipping seedlings for 30 min had 0.019 mg/m^3 permethrin in the breathing zone. A gas-chromatographic analysis with a detection limit of 0.1 µg/ml found no acid metabolite moiety in his urine. The concentrations in the breathing area of workers handling the treated seedlings was 0.011 to 0.085 mg/m^3 of permethrin. After 6 h of exposure the highest value in urine was 0.26 µg/ml of acid metabolite moiety. The concentrations of permethrin when planting the treated seedlings was 0.001 to 0.006 mg/m^3. No acid metabolite was determined in the urine of the planters.

Exposure of the general population is possible mainly during home desinsection and food consumption. Some residues found in milk and grains are shown in Table 4; these residues are negligible compared to ADI. Stored grains have higher concentrations, which diminish 10-fold after milling and baking.

The most important exposure for the general population is the use of SPs for home and personal desinsection purposes. Thus Nassif et al. determined a permethrin metabolite in urine after applying powder inside the clothes of 15 persons.[27] The concentrations in the urine after 24 h were 0.12 µg/ml and 0.35 µg/ml respectively. No excretion was found after 30 d.

TABLE 4
Residues of Synthetic Pyrethroids[18,20,26]

Kind of SP	Milk, butter	Grains (stored)	Bread
		Levels in food	
Delthamethrin	0.00125 mg/kg 0.0255 mg/kg	1.08 mg/kg 9 months after treatment	0.11 mg/kg
Permethrin	0.01–0.3 µg/g (feeding 0.2 ppm and 100 ppm)	0.36–4.5 mg/kg (0.9 mg/kg 9 months after treatment)	0.12 mg/kg
Fenvalerate	0.02 mg/kg after spraying	0.74 mg/kg 10 months after treatment	0.06–0.1 mg/kg
d-Phenothrin	—	3.5–3.84 mg/kg (6 months after treatment)	0.66–0.69 mg/kg
Allethrin	0.1 ppm	—	—
Resmethrin	—	1.1–1.9 mg/kg (0.4 mg/kg 6 months after treatment)	0

VII. EFFECTS ON HUMANS

In humans, dermal exposure of deltamethrin, fenvalerate, permethrin, and cypermethrin causes transient itching and burning sensations in the face, without clinical signs of inflammation. This paraesthesia may result in discomfort and anxiety. There are only a few reports on such effects due to occupational exposure to pyrethroids.

Pyrethroids are neurotoxic compounds. Severe reactions have not been reported in man; in mammals, however, increased activity, loss of coordination, paralysis, and death are reported.

In order to assess the significance of these symptoms, Quesne et al. studied 23 workers, exposed to different levels of synthetic pyrethroids; 19 of these workers had experienced 1 to 10 or more episodes of sensory facial irritation after being exposed to fenpropathrin, cypermethrin, and fenvalerate.[28] The sensation was described as tingling, burning "like coming in from the cold", and "nettle-rash". It developed within 30 min after beginning work and up to 3 h after the exposure ceased; it persisted from 30 min up to 8 h. The subjects were not aware of the loss of sensation when they touched or rubbed the affected area. The occurrence and severity of the symptoms could be related to the degree of exposure. It is also possible that symptoms developed more easily after repeated exposure.

It seems that direct skin contact with pyrethroid is necessary for the development of symptoms. Neurological examination detected no abnormalities. The electrophysiological thorough checking of the arms and legs of each worker showed no significant differences in comparison with control groups of unexposed workers. The authors suppose that the sensory irritation (paraesthesia, dysaesthesia) might be due to the spontaneous firing of sensory nerve fibres or the transiently lowered threshold of sensory nerve endings by the pyrethroids.

Kolmodin-Hedman et al. questioned 139 persons working in the plant nursery.[17,29] The symptoms were mainly irritative, such as itching and burning of the skin, itching and irritation of the eyes, and allergic reactions. Persons using fenvalerate and permethrin as a wettable powder reported coughing, dispnea, sneezing, and increased nasal secretion. Irritative symptoms were reported in 73% of the workers for fenvalerate, 63% for permethrins *trans/cis* 75/25 wettable powders, and 33% for permethrins *trans/cis* 60/70 (emulsion in organic solvents). Higher rates were reported in planting. The authors held the dried insecticide that fell off the conifer needles accountable. Planting sticks clearly reduced the frequency of subjective symptoms according to Edling et al.[29]

Tucker and Flannigan studied the cutaneous effect of occupational fenvalerate exposure in agricultural application.[30] All the workers (pilots, mixer-loaders, sprayers — farm machinery operators, a total of 16 subjects) noted a cutaneous sensation, varying according to the exposure. The sensation was described as stinging or burning, which progressed to numbness in approximately one fourth of the workers. One third complained of sleeping disturbances (insomnia).

Paraesthesia was noted by some individuals within minutes after exposure, but symptoms developed some hours later. Sweating, heat, sunlight, and washing the exposed areas with water enhanced the sensation. The greatest exacerbation occured at the time of sleep. All symptoms disappeared after a 24 h period. Clinical signs of inflammation, such as edema or vesiculation, were absent. The authors suggest that acid metabolite moiety in the urine is of little practical value. They concluded that these effects were not dependent on race or personal history of atopy.

The skin reaction is different only for deltamethrin. Tenacious and painful pruritus, followed by blotchy local burning and 2 d of erythema, as well as desquamation, are observed. Rhinorrhea and lacrimation are common.

Tomova[31] and Bainova[32] reported cases of allergic contact dermatitis among greenhouse workers exposed to deltamethrin. The patch test with 0.05% decis gave positive results. Bainova et al. observed paresthesia among 25 workers formulating deltamethrin for domestic use (0.02, 0.04, and 0.08%).[33] The dysesthesia was concentration-related and increased during washing with soap or organic solvents. Recovery occurred after 3 to 12 h. Adaptation to the sensory reaction did not occur. After several months of exposure, lacrimation, coughing, dispnea, and skin eruption were demonstrated. Higher doses produce transient damage to the nervous system, such as axonal swelling and breaks and myelin degradation in sciatic nerves. Delayed neurotoxicity was not demonstrated.

All synthetic pyrethroids are allergens — an effect very well known for natural pyrethrin. They provoke slight to moderate contact sensitization in guinea pigs, which correlates with the data from epidemiological studies in agricultural workers.[13,17,29] On the other hand, they affect the immune system. Thus, Desi et al. found both the humoral and cell mediated immune response of rabbits and rats suppressed due to cypermethrin treatment.[14]

Respective investigations on test systems *in vitro* and *in vivo* have proven that pyrethroids do not have mutagenic, gonadotoxic, embryotoxic, teratogenic and carcinogenic effects. Metabolites of some pyrethroids (Bα-isomers) form lipophilic conjugates such as cholesterol esters, which are persistent mainly in fat tissues. Only Bα-isomers produce microgranulomatous lesions in rats and mice. Kaneko et al. suggest a relationship between granuloma formation and this kind of conjugation.[15]

Bainova and Kaloyanova reported that the organic solvents and surfactants in commercial products increase the dermal penetration and irritative potency of pyrethroids, especially the alpha-cyano derivatives.[16]

A study on volunteers with d-phenothrin was performed with 32 mg/man (0.67 to 0.44 mg/kg/b.w.) dermally.[48] No significant skin abnormalities were found nor any changes in blood biochemistry and morphology.[18] Study on the cutaneous exposure of volunteers to fenvalerate was reported by Knox et al.[34] The onset of cutaneous sensation of numbness, itching, burning, and warmth occured in 1 h peaked at 3 to 6 h and disappeared after 24 h.

Primary irritative contact dermatitis from fenvalerate and permethrin were reported in experimental studies on albino rabbits by Flannigan et al.[35] Slight erythema was noted visually, which correlated with the increased cutaneous blood flow measured by laser Doppler velocimetry. These minor changes could not suggest that clinical symptoms of irritation would be significant. This suggestion was supported in another study on volunteers, performed by

Flannigan et al.[36] A comparative study on paraesthesia occurence caused by fenvalerate and permethrin was performed on volunteers. Technical and formulated products were applied to the earlobe at a dose of 0.13 mg/cm^3. A linear correlation between the concentration of pesticide and the degree of induced dysaesthesia was observed for both pyrethroids. This cutaneous sensation was not related to the sensitivity of persons and seems to be almost universal.

Fenvalerate produced a fourfold increased paraesthesia in comparison with permethrin.

Flannigan and Tucker studied the effect of flucythrinate on skin in volunteers.[37] Technical grade flucythrinate was applied to the forearms of human participants twice daily for five consecutive days. Paraesthesia was noted 5 h after the initial application. It was described as a burning to tingling sensation that eventually subsided to a mild degree of pruritus by day three. No signs of inflammation (erytheme, edema, or vesiculation) were reported. The pyrethroid did not inhibit the histamine induced axon reflex vasodilation.

An effective treatment is not known. Diazepam and phenobarbital are used, but not always effectively. Natrium bicarbonicum is used for stomach lavage if swallowed. Atropin can depress the salivation and other cholinergic symptoms. The skin and mucous should be carefully washed.

For skin effects a highly efficient therapeutic agent for pyrethroid exposure is, according to Flannigan et al., tocopherol acetate (vitamin E acetate).[36] Topical application of vitamin E acetate successfully inhibited paraesthesia in almost 100% of cases.

Vitamin E acetate acts as a biological antioxidant that inhibits peroxides from accumulating and protects cells. It contributes to the stability and integrity of biological membranes, including lysosomal membranes. It thus prevents inflammatory processes, histamine liberation from the granules of most cells, and serotonin liberation from tissue cells. Flannigan et al. suggests vitamin E acetate for protection against occupational exposure in the field.[36]

VIII. PREVENTION

It is generally accepted that the SP compounds provide an adequate margin of safety when used in the prescribed manner for desinsection.[37]

Water flowable concentrations are less dangerous for skin effects. It is not recommended that persons with atopic allergy or skin or respiratory tract sensibilization to chemicals should work with pyrethroids.

Protecting the skin and respiratory system is necessary.

The high frequency of irritative symptoms of skin and respiratory tract in connection with fenvalerate led to the withdrawal of this product from the market for use in forestry in Sweden.[29]

Sasinovitch and Panshina[38] calculated the maximum allowable concentrations for some SPs on the basis of toxicometric data, i.e:

Decametrin 0.1 mg/m3
Fenvalerate 0.2 mg/m3
Cypermethrin 0.3 mg/m3
Permethrin 0.5 mg/m3

Acceptable daily intake for some SP is shown below.

Kind of SP	Mg/kg/day
d-Phenotrin	0–0.04
Deltametrin	0–0.01
Cypermethrin	0–0.02
Cyhalothrin	0–0.02
Fenvalerate	0–0.2
Permethrin	0–0.05

IX. CONCLUSION

Synthetic pyrethroids are insecticides of high activity. They are widely used in agriculture and in-home desinsection.

In the organism they are easily metabolized and do not accumulate in tissues.

Pyrethroids are neurotoxic compounds. In humans, intoxications are usually very light, and neurological symptoms, such as generalized tremor and seizures, are found only in animals.

Sensory facial irritation is a very common syndrome (paraesthesia and dysaesthesia).

Pyrethroids with alphacyanogroup demonstrated some symptoms similar to those of organophosphorous intoxication, such as salivation, myosis, bradycardia, and tremor of the extremities.

Dermal irritative and allergic effects are very often reported following occupational exposure.

REFERENCES

1. **Roberts, T. R. and Standen, M. E.,** Degradation of the pyrethroid cypermethrin NRDC 149 (±)-α-cyano 3-phenoxybenzyl(±)-cis, trans-3-(2,2-dichlorvinylcyclopropanecarboxylate) and the respective cis-(WRDC 160).
2. **Kuzo, L. O. and Casida, J. E.,** Metabolism and toxicity of pyrethroids with dihalovinyl substituants, *Environ. Health Perspect.,* 21, 285, 1977.
3. **Carlson, G. P. and Schoonig, G. P.,** Induction of liver microsomal NADPH cytochrome C reductase and cytochrome P-450 by some new synthetic pyrethroids, *Toxicol. Appl. Pharmacol.,* 52, 507, 1980.
4. **Eadsforth, C. V. and Baldwin, M. K.,** Human dose-excretion studies with the pyrethroid insecticide cypermethrin, *Xenobiotica,* 13, 67, 1983.
5. **Worthing, C. R.,** Ed., *The Pesticide Manual. A World Compendium,* English edition, The British Crop Protection Council, 1988.
6. **Soderlund, D. M., Chiasuddin, S. M., and Helmeth, D. W.,** Receptorlike stereospecific binding of a pyrethroid insecticide to mouse brain membraines, *Life Sci.,* 33, 261, 1983.
7. **Vijerberg, H. P. and van der Berken, J.,** Frequency-dependent effects of the pyrethroid insecticide decamethrin in frog, *Eur. J. Pharmac.,* 58, 501, 1979.
8. **Staatz, C., Bloom, A. C., and Lech, J. J.,** Effect of pyrethroids on ^{3}H kanic acid binding to mouse forebrain membranes, *Toxicol. Appl. Pharmacol.,* 64, 566, 1982.
9. **Staatz, C. G., Bloom, A. S., and Lech, Y. I.,** A pharmacological study of pyrethroid neurotoxicity in mice, *Pest. Biochem. Phys.,* 17, 284, 1982.
10. **Verschoyle, R. D. and Aldridge, W. N.,** Structure-activity relationships of some pyrethroids in rats, *Arch. Toxicol.,* 45, 325, 1980.
11. **Staatz, C. G. and Hosko, M. J.,** Effect of pyrethroid insecticides on EEG activity in conscious immobilized rats, *Pest. Biochem. Phys.,* 24, 231, 1985.
12. **Brodie, M. E. and Aldridge, W. N.,** Elevated cerebellar cyclic GMP levels during the deltamethrin-induced motor syndrome, *Neurobehav. Toxicol. Teratol.,* 4, 109, 1982.
13. **Bainova, A.,** Toxicological Problems Related to the Action of Chemicals on the Skin (Dermatotoxicology), thesis Dr. So, Sofia, 1985.
14. **Desi, I., Varga, L., Dobrony, I., and Szkolenic, G.,** Immunotoxicological investigation of the effects of a pesticide; cypermethrin., *Arch. Toxicol. Suppl.,* 8, 305, 1985.
15. **Kaneko, A., Masatoshi, M., and Inushi, M.,** Differential metabolism of fenvalerate and granuloma formation. I. Identification of a cholisterol ester derived from a specific chiral isomer of fenvalerate, *Toxicol. Appl. Pharmacol.,* 83, 1, 148, 1986.
16. **Bainova, A. and Kaloyanova, F.,** Study of the skin allergic and irritative effects of synthetic pyrethroids, *Hig. Zdraveop.,* 28, 2, 19, 1985.
17. **Kolmodin-Hedman, B., Swenson, A., and Akerblom, M.,** Occupational exposure to some synthetic pyrethroids (permethrin and fenvalerate), *Arch. Toxicol.,* 50, 1, 27, 1982.
18. FAO/WHO, 1975 Evaluation of Some Pesticide Residues in Food, World Health Organization, Geneva, 1976, 17.

19. FAO/WHO, 1976 Evaluation of Some Pesticide Residues in Food, World Health Organization, Geneva, 1977, 55.
20. FAO/WHO, 1979 Pesticide residues in food. 1978 evaluating, Rome, FAO of UN, p 427.
21. FAO/WHO, Pesticide residues in food. 1979 Evaluations. FAO, Rome, 299, 1980.
22. FAO/WHO, Pesticide residues in food. 1981 evaluations, FAO, Rome, 209, 1982.
23. FAO/WHO, 1981 Evaluation of Some Pesticide Residues in Food, World Health Organization, Geneva, 1982, 113.
24. FAO/WHO, 1982 Evaluation of Some Pesticide Residues in Food, (FAO Plant Production and Protection Paper. 49), FAO, Rome, 1983.
25. FAO/WHO, Pesticide residues in food. 1984 Evaluation, FAO, Rome, 63, 1984.
26. WHO, Technical Report Series 1973, 513, 1973.
27. **Nassif, M., Brooke, J. P., Hutchinson, D. B. A., Kamel, O. M., and Savage, E.,** Studies with permethrin against bodylice in Egypt, *Pest. Sci.,* 11, 679, 1980.
28. **Le Quesne, P. M., Maxwell, U. C., and Butterworth, S. T.,** Transient facial sensory symptoms following exposure to synthetic pyrethroids. A clinical and physiological assessment, *Neurotoxicol.,* 2, 1, 1980.
29. **Edling, C., Kolmodin-Hedman, B., Akerblom, M., Rand, G., and Fischer, T.,** New methods for applying synthetic pyrethroids when planting conifer seedlings. Symptoms and exposure relationships, *Ann. Occup. Hyg.,* 29, 421, 1985.
30. **Tucker, S. B. and Flannigan, S. A.,** Cutaneous effects from occupational exposure to fenvalerate, *Arch. Toxicol.,* 54, 195, 1983.
31. **Tomova, L. M.,** Study of occupational dermatoses among field workers, Referat (Ph.D. thesis), Sofia, 1982.
32. **Bainova, A.,** Synthetic pyrethroids — a new group of plant protective drugs, *Savr. Med.,* 38, 7, 3, 1987.
33. **Bainova, A., Mihovski, M., Ismirova, N., and Benchev, I.,** Specific dermal irritation after exposure to synthetic pyrethroid, 4th Congress of Dermatology, Varna, 1986, 2.
34. **Knox, J. M., Tucker, S. B., and Flannigan, S. A.,** Paraesthesia from cutaneous exposure to a synthetic pyrethroid insecticide, *Arch. Dermatol.,* 120, 744, 1984.
35. **Flannigan, S. A., Tucker, S. B., Key, M., Ross, E., Fairchild, J., Grimes, A., and Harrist, B.,** Primary irritant contact dermatitis from synthetic pyrethroid insecticide exposure, *Arch. Toxicol.,* 56, 288, 1985.
36. **Flannigan, S. A., Tucker, S. B., Key, M. M., Ross, C. E., and Fairchild, R. G.,** Synthetic pyrethroid insecticides. A dermatological evaluation, *Br. J. Ind. Med.,* 42, 6, 363, 1985.
37. **Flannigan, S. A. and Tucker, S. B.,** Variation in cutaneous perusion due to synthetic pyrethroid exposure, *Br. J. Ind. Med.,* 42, 773, 1985.
38. **Sasinovich, L. M. and Panshina, T. N.,** Obosnovanie gigienicheskih reglamentov soderzjaniya sinteticheskih piretroidov v vozdyuhe rabochej zony, *Gig. Truda Prof. Zabol.,* 8, 48, 1987.
39. IRPTC Data profile, International Register of Potentially Toxic Chemicals, legal file.

Chapter 6

ORGANOTIN COMPOUNDS

I. INTRODUCTION

The development and use of organotin compounds as pesticides is a comparatively recent process. They have a wide spectrum of effects and are applied for different purposes. Mostly three substituted organotin compounds are used as pesticides: trialkyl (mostly tributyl) tin compounds, triphenyl- and tricyclohexyltin compounds.

In the last years, concern has been expressed about the future application of organotin (particularly tributyltin) compounds, in view of their ability to affect some marine organisms and induce malformations in oysters. Organotin compounds find application as antifoulants in boat paints and fish farm nets and cages.[1-3] Due to teratogenicity in mammals, the use and production of plictran was discontinued by the firm producer. There is no data on the long-term effects on human beings. Existing data on human toxicity and clinical effects are scarce.

II. PROPERTIES

The properties of organotin pesticides differ in the individual compounds. All are strong oxidants. The properties of some of the most used organotin compounds are shown on Table 1.[4]

III. USES

The uses of organotin compounds in agriculture are summarized in Table 2.

Fentin hydroxide and acetate are used as fungicides in doses of 250 to 400 g/ha and 160 to 260 g/ha, respectively.

IV. METABOLISM

Organotin compounds enter organisms by inhalation, via skin, and orally. The highest concentrations have been detected in the liver and brain.

In the liver, as well as in some other organs and tissues, organotin compounds are biotransformed through dealkylation and dearylation by the microsomal mono-oxygenase system. They are excreted in urine, but some can be eliminated with bile and feces. The half-life in different organs is 3 to 40 d.

V. TOXICITY: MECHANISM OF ACTION

Oral LD_{50} for trialkyltin compounds is strongly influenced by the length of the alkyl chain. Thus, LD_{50} of methyltin compounds is 9 mg/kg, while LD_{50} of N-octyltin is more than 1000 mg/kg.[5] LD_{50} of fenbutantin-oxide in rats is 2631 mg/kg.

Dermal LD_{50} of trimethyltin compounds is 50 to 100 mg/kg (mice), and for bis(tributyltin)oxide, it is about 200 mg/kg (rats and mice). Contrary to the high dermal toxicity of alkyltin compounds, triphenyltin acetate does not penetrate the skin.[6]

Tributyltin compounds are strong skin irritants in humans especially if the site of action is the depths of hair follicles.[7] The most active are di- and trialkyltin compounds. Dialkyltin causes generalized illness, with damage to the biliary tract a main pathological feature (dermal application). Trialkyltin compounds act on the central nervous system and the majority of them

TABLE 1
Properties of Organotin Pesticides

Structural formula	Names and properties
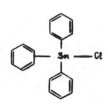	**Fentin acetate (Bretdan)** Colorless crystal solid, m.p. 121, v.p 1.9 m Pa Solubility in water 9 mg/l (20°C)
	Fentinhydroxide (Duter phenostat-H) Colorless crystalline solid, m.p. 116 v.p 0.047 m Pa Solubility in water 1 mg/l (20°C)
	Triphenyltinchloride (aquatin, phenostat C, tinmate) Colorless crystal, m.p.105°C insoluble in most organic solvents
	Fenbutatin oxide, Osadan, Torgue, Vendex White colorless solid Insoluble in water
	Azocyclotin (perepal) Colorless crystal m.p. 218 v.p 5 × 10⁻⁵ m bar (20°C) Solubility 0.001 mg/l at 20°C

produce interstitial edema. Triethyltin is the most active of the trialkyl compounds.[5] Trimethyltin and triethyltin compounds are potent inhibitors of oxidative phosphorylation in the mitochondria (Aldrige, 1976, cited by WHO/EHC[7]). Triorganotin compounds impair mitochondrial function in three different ways: they discharge a hydroxylchloride gradient across mitochondrial membranes, they interact with the basic energy system involved in ACTP synthesis, and they interact with mitochondrial membranes causing swelling and disruption.[7] The antagonistic action of dimercaptol and diethyltin suggests that it acts through combination with SH-groups.[8]

VI. EXPOSURE: DOSE-EFFECT RELATIONSHIP

Organotin compounds used as molluscicides or antifouling paints may be present in water.[10] Organotin concentrations were found in harbor water, the highest values being 0.88 μg/l. A bioaccumulation factor up to 10,000 in oysters is reported for tributyltin oxide.[9]

TABLE 2
Organotin Pesticides and Their Specific Use

Use	Name
Fungicides in agriculture and paper and paints manufacturing	Triphenyltin acetate triphenyl hydroxide
Bactericides and disinfectants in hospitals	Tributyl benzoate
Helminticides in poultry	Dibutyltin dilaurate tetraisobutyltin
Nematocides	P-biomophenoxy triethyltin
Herbicides	Trivinyltin chloride
Rodent repellents	Tributyltin chloride, tributyltin chloride and acetate
Molluscicides, ovicides, chemosterilisants	Triphenyl- and tributyltin compounds
Acaricides	Tricyclohexyltin hydroxide (plictran)
General biocides	Bis(tributyltin) oxide

Occupational exposure is possible during field application of organotin pesticides.

In food, organotin compounds are found as follows: tricyclohexyltin hydroxide up to 2 mg/kg in fruit (apples and pears) on the day of final application (4 treatments in total); triphenyltin hydroxide, acetate, and chloride up to 0.1 mg/kg in vegetables; and triphenyltin actate 0.004 mg/l in cow milk.[7] Cooking does not destroy those compounds.

VII. EFFECTS ON HUMANS

A. DERMAL CONTACT

Two types of skin lesions are observed when dermal contamination occurs: acute local burn and subacute dermatosis. Tributyltin produces skin burn on the unclothed part of the body when the compound is in contact with the skin for more than a few minutes.[11] A more diffuse type of lesion is produced by prolonged contact with moistening by vapor or liquid.

Lyle produced acute lesions on the backs of the hands of 5 volunteers with tributyltin monochloride.[11] Visible reaction appeared after 2 to 3 h — reddening and swelling of the mouths of the hair follicles. After 8 h follicular inflammation progressed. The skin between the follicles was only slightly affected, and there was only a minor degree of edema. On the second day sterile small pustules developed over the follicular openings. They dried up on the third or the fourth day. Healing was complete within 10 d. Pruritus was confined to the tested area and persisted for no more than 3 d after application.

Subacute lesions are characterized by itching — erythematous eruption with well expressed demarkation from normal skin. The affected skin is slightly sticky and abnormally friable; scratch marks are usually present. No pustules or folliculitis are observed, as they are in acute local burns. After an involuntary splashing of the eyes, lacrimation and intense suffusion of the conjunctivae appeared within minutes, despite immediate lavage, and persisted for 4 d.

Lyle made a comparative study of the irritation potential of different butyltin compounds.[11] Chemical burns were produced by tributyltin chloride, actate, laurate, and oxide. From dibutyltin compounds only dibutyltin chloride demonstrated skin irritation.

Dermal contact (contamination on hands and chest) with 60% triphenyltin, in combination with 15% maneb, produced skin injuries and systematic symptoms. General weakness, headache, nausea, and epigastric pain occurred. Liver transaminases (SGPT) were elevated during 2 months. The liver was painful, tender and enlarged. Chronic hepatitis developed and persisted for 3 years.[12]

B. INHALATORY EXPOSURE

In workers, symptoms of respiratory irritation such as sore throat and cough, occuring several hours after exposure to vapor or fumes, are reported.[11]

Two pilots exposed to fungicide mixtures during work had gastric pain, diarrhea, dryness of the mouth, severe thirst, and vision disturbances. Toxic hepatitis with hepatomegalia developed. Transaminase was elevated, and the highest values appeared after an interval of 6 weeks. Liver biopsy demonstrated diffuse steatosis. Hyperglycemia and glucosuria were present. Recovery occurred after 1 year in 1 of the pilots and after 6 weeks in the other. Three mechanics were also involved with less serious symptoms: diarrhea, headache, eye pains, blurred vision, epigastric pain, and thirst. Exposure was mostly inhalatory, but possible ingestion with contaminated food could not be excluded.[14]

Workers with long-term exposure to butyltin compounds, because of their involvement in its production, reported reduced sense of smell, headache, nasal hemorrhages, lassitude, and a feeling of stiffness in the shoulders (Akatsuka et al., cited by WHO/EHC[7]).

Markećević and Turka have reported the effects on the respiratory system.[13] Workers engaged in triphenyltin formulation developed conjunctival and upper respiratory tract irritation. The skin on the hands and scrotum was also affected.

C. ORAL EXPOSURE

Although inhalatory and dermal exposure do not reportedly produce severe cerebral edema and death, ingestion of diethyltin diodide dissolved in linoleic acid (used for treatment of furunculosis) produced 210 intoxications with 110 deaths.[8] The symptoms were cerebral edema, severe headache, nausea, vomiting, visual disturbances, and blindness in 33% of the cases. Ophthalmoscopy discovered papilledema and papillary stasis. Psychological disturbances, meningeal irritation, and other central nervous system symptoms were reported as well.[7]

VIII. TREATMENT

There is no sufficient information concerning the treatment of organotin intoxication. Dimercaptol seems to have protective effects against some dialkyltin compounds reacting with sulfhydryl groups, but not against trialkyltin compounds.

Symptomatic therapy to diminish the severity of brain edema and even surgical decompression are suggested. (Alajeranine et al., cited by WHO/EHC[5])

Studer et al. studied the effect of dexamethazone (steroid therapeutic) on triethyltin bromide-induced cerebral edema.[15] The result was decreased mortality and severity of brain edema. According to the authors, this was due to enhanced excretion or catabolism of triethyltin bromide.

IX. PREVENTION

Safe levels for all organotin compounds are as follows: TLV in U.S. is calculated at the tin basis of 0.1 mg/m³; LDLH is 200 mg/m³.[16]

ADIs recommended by FAO/WHO are as follows: Azocyclosin — 0.003 mg/kg/b.w./d, fentin hydroxide, acetate and chloride — 0.0005 mg/kg/b.w./d.[17]

In the U.S. a triphenyltin dose of 0.0003 mg/kg/b.w./d is accepted as ADI. Concentrations of organotin compounds in drinking water should not exceed 0.1 µg/l.[18]

X. CONCLUSION

Organotin compounds are used as pesticides with a wide spectrum of effects. Their toxicity differs in relation to their structure — very high for methyltin compound and low for fenbutatin oxide. Some of them are teratogens in animals.

Most of the organotin compounds are strong irritants. Dermal contamination produces burns and subacute dermatitis. Inhalatory exposures produce respiratory irritation. Toxic hepatitis, hepatomegalia, and cerebral edema may occur with high inhalatory exposure or oral intake.

REFERENCES

1. **Langhlin, R. B. and Linden, O.,** Fate and effects of organotin compounds, *Ambio,* 14, 2, 88, 1985.
2. **Beaumont, A. R. and Newman, P. B.,** Low levels of tributyl tin reduce growth of marine micro-algae, *Mar. Pollut. Bull.,* 14, 10, 457, 1986.
3. Tributyl tin should be banned without delay, Press release, Fields of the Earth, 53 George IV Bridge, Edinburgh, Scotland, February 6, 1987.
4. **Worting, C. R. and Walkers, B.,** The pesticides manual. A world compendium, 8th Edition, The British Crop Prot. Council.
5. **Barnes, J. M. and Stoner, J. B.,** Toxic properties of some dialkyl and trialkyl tin salts, *Br. J. Ind. Med.,* 15, 15, 1958.
6. **Stoner, H. B.,** Toxicity of tripenyltin, *Br. J. Ind. Med.,* 23, 22, 1966.
7. **Anon.,** Tin and organotin compounds. A preliminary review, Environmental Health Criteria Document/World Health Organization, 15, World Health Organization, Geneva, 1980.
8. **Stoner, H. B., Barnes, J. M., and Duff, J. I.,** Studies on the toxicity of alkyl tin compounds, *Br. J. Pharmacol.,* 10, 16, 1955.
9. **Waldock, M. J. and Thain, J. E.,** Shell thickening in Crassostrea gigas; Organotin antifouling or sediment induced, *Mar. Pollut. Bull.,* 14, 11, 411, 1983.
10. **Cleary, J. J. and Stebbing, A. R. D.,** Organotin and total tin in coastal waters of South-West England, *Mar. Pollut. Bull.,* 16, 9, 351, 1985.
11. **Lyle, W. H.,** Lesions of the skin in process workers caused by contact with butyltin compounds, *Br. J. Ind. Med.,* 15, 193, 1958.
12. **Mijatovic, M.,** Hepatitischronica kao posledica ekspozicije "Brestanu", *Jug. Inostr. Dok., Zast. Radu,* 8, 8, 3, 1972.
13. **Markećević, A. and Turko, V.,** Lesions due to tripenyltin acetate "Bresdan", *Arch. Hig. Rada,* 18, 355, 1967, (in Serbo-Croatian).
14. **Horacek, V. and Demcik, K.,** Skupinova otrava pri postriku polnich kultur Brestanem-60 (oktanem triphenylcinicitym), *Prac. Lek.,* 22, 2, 61, 1970.
15. **Studer, R. K., Siegel, B. A., Morgan, J., and Potchen, E. J.,** Dexamethazone therapy of triethyltin induced cerebral edema, *Exp. Neurol.,* 38, 429, 1973.
16. National Institute for Occupational Safety and Health (U.S.A.) Pocket Guide to Chemical Hazards, U.S. Department of Health and Human Services, September 1985.
17. Pesticide reference: Index, JMPRS-IARC-IPCS-IRPTC-BC, Vettorasi, G., Ed., IPCS, 1961–1983.
18. **Anon.,** Organotin compounds, Resummé 5801, (Unpublished material), World Health Organization/EURO.

Chapter 7

ORGANOMERCURIAL COMPOUNDS

I. INTRODUCTION

One of the oldest groups of fungicides used for seed dressing are organomercurials. The most used compounds from this group are methyl, ethyl, methoxyethy, and phenyl mercury compounds.

In the environment soil aerobic microorganisms subject organomercurial compounds to reduction to alifatic or aromatic hydrocarbons and mercury. In soil mercury is present as an oxide or sulfide. Some plants can absorb some of the mercury compounds from the soil.[1]

Mercury is accumulated in the tissues of mammals and thus can go through the trophic chain into humans. This is the main reason for the considerably decreased use of organomercurial compounds as fungicides for seed dressing. Another very important reason is the misuse of treated corn for bread or animal food.

II. PROPERTIES

Organomercurials have different volatilities. Ethylmercuric chloride is rather volatile. At usual room temperature it can be emitted in air. At 20°C its volatility is 12 to 21 mg/m^3, and its vapor pressure is 8.4×10^{-4} mg Hg, compared to 0.9×10^{-6} mg Hg at 30°C for the less volatile phenylmercuryacetate. The chemical structure and some properties of the most used organomercurial fungicides are shown in Table 1.

III. USES

Treatment of seeds by organomercurial fungicides is performed in special hermetic equipment in the open air or in well ventilated premises.

The sowing of treated seeds is performed by a seed drill specially adapted for this purpose.

IV. METABOLISM

Organomercurial fungicides enter the body through respiratory or gastrointestinal systems; they can also penetrate the skin.

Organomercurials accumulate in body tissues, mainly in the liver, kidney, brain, etc. They are excreted very slowly by the kidneys and gastrointestinal system. They also can be found in breast milk.

V. TOXICITY: MECHANISM OF ACTION

Organomercurial fungicides possess very high acute and chronic toxicity. LD_{50} for different compounds varies from 30 (alkylmercury chloride) to 80 (phenylmercuryacetate) mg/kg/b.w.

The nervous system and kidney are the main targets of their toxicity, but the cardiovascular system, hemopoiesis, and liver are also affected.

The effect of organomercurial compounds on enzymes are similar to that of mercury. They are known as thiol poisons. Entering in reaction with the SH-group of the proteins, they disturb the activity of enzymes.

TABLE 1
Properties of Some Organomercurial Pesticides

Structural formula	Names and properties
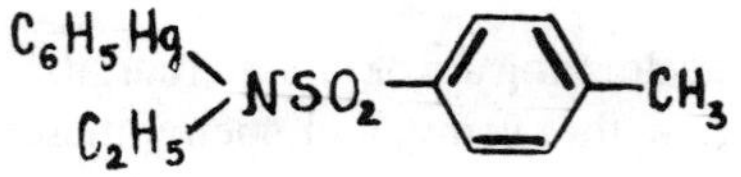 C_6H_5Hg, C_2H_5 $>NSO_2$—⟨benzene ring⟩—CH_3	Ethylmercury, tolyol sulfonilide Mol. mass 475, m.p. 156°C Insoluble in water
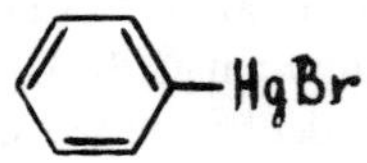 ⟨benzene ring⟩—$HgBr$	Phenylmercurybromide Mol. mass 357, m.p. 275°C Unsoluble in water
$C_6H_5HgOCOCH_3$	Phenylmercuryacetat Mol. mass 336, m.p. 150°C Solubility 24.7 g/l
$C_2H_5HgC\ell$	Ethylmercurychloride Mol. mass 265, m.p. 192°C Slightly soluble

Pregnant women are more sensitive to organomercurial compounds. An exposed mother risks malformations in the newborn. Since mercury can be excreted by breast milk, maternal exposure may be dangerous for the newborn.[2] Organomercurial fungicides are allergens.

VI. EXPOSURE: DOSE-EFFECT RELATIONSHIP

At seed dressing the concentrations of organomercurial fungicides may reach 0.12 to 0.6 mg/m^3. At the storage place the concentrations can be 0.01 to 0.03 mg/m^3. At transportation, seeds dressed with organomercurials can create high concentrations in the breathing zone, ranging from 0.157 to 0.85 mg/m^3; at sowing the range is from 0.095 to 0.2 mg/m^3.[1]

Exposing the population to mercury by using organomercurial fungicides in rice growing is an important problem in Japan.[3] Mercury levels in the hair of Tokyo residents is 2 to 3 mg/kg; in the rest of the Japanese population it is about 6 mg/kg, or 3 times higher than the values in the population of Europe. Mercury content in the urine of the Japanese population is 2 times higher than in the Europeans.

Wakatsuki has shown that the hair of people living in farms contains more mercury than the hair of town inhabitants.[4]

Mercury passes through the placenta and accumulates in the bones of the fetus. According to Wakatsuki, the mercury content (in mg/kg) for urban and rural dwellers, respectively, is 5.60 and 7.99 in the mother's hair, 6.67 and 11.08 in the newborn's hair, 0.26 and 0.60 in the placenta, and 0.55 and 0.68 in mother's milk. According to the author, this accumulation may be a cause of latent health disturbances.

Similar data are reported by Kassai.[5] The mercury content in hair has been found to vary between 1.04 and 23.39 mg/kg, with an average value of 7.44 mg/kg. In rice it has been found to be about 0.06 mg/kg. Natural mercury exerts a possible influence; it accumulates in river fish as rivers sweep along mercury from the rocks and soils. Sometimes up to 0.5 mg/kg of mercury is in fish and the zone of Minamata contains up to 35 mg/kg.

In Japan, legislative measures are being taken for the limitation and ban of organomercurial preparations in agriculture.

By the method of the flameless atom absorption spectrophotometry, Juszkievicz et al. have determined the mercury concentrations in samples of breast milk obtained from 101 mothers 4 d after delivery.[6] From 0.001 to 0.01 mg/kg mercury was found in all samples, with a mean concentration of 0.0059 mg/kg. Higher concentrations were found in women over 30 years old. In women exposed to organomercurial fungicides, mercury reached up to 0.04 mg/l.

Juszkievicz and Sprengier investigated the mercury content in the kidney tissue of horses aged from 5 to 20 years and raised in 16 different regions.[7] The mean mercury concentration was 0.11 mg/kg, which was considered normal. On the basis of these results the authors concluded that the use of organomercurial fungicides in agriculture does not influence substantially the accumulation of mercury in kidney tissue.

Sprengier determined the mercury content in the livers and kidneys of cats from two regions — agricultural and industrial.[8] The values for the cats of the industrial region proved to be significantly higher than those for the cats of the agricultural region.

VII. EFFECTS ON HUMANS

The toxicity of organomercurial compounds depend on their composition. Methylmercurial compounds are especially toxic for the CNS. Acute intoxications are very rare and manifested as a toxic encephalopathy with noncommittant gastrointestinal and renal disturbances. The main symptoms are headache, nausea, vomiting, metallic taste, salivation, abdominal pains, diarrhea, tremor, paralysis, joint aches, vision and hearing disturbances, coma, proteinuria, and leukocytosis.

The early clinical symptoms of intoxication are weariness, general weakness, unnatural taste in the mouth, gum hemorrhage, and sleep disturbances. Sometimes visual and auditory hallucinations appear, along with increased irritability, feeling of fear, apathy, and tremor. Thanks to the use of low volatility compounds, occupational intoxications are rare. Mainly occupational dermatites have been reported. Visual impairments have been reported by Medved et al., optic neuritis — by Bidstrup, atrophic alterations in fundus oculi — by Katzunuma et al., allergic manifestations — by Patty, and changes in electrolyte metabolism and cardiovascular system — by Miatskanov et al.[9-13]

Taylor et al.[14] published their observations on 33 subjects engaged in dressing corn with organomercurial compounds. The workers were followed up from 2 months to 3 years after exposure. Proteinuria and mercury in urine, glucophosphatisomerase inhibition in serum, and increased mercury content in blood were found. Several nonoccupational mass poisonings by organomercurial fungicides are known. Of a most instructive among them is a mass intoxication by corn, treated with methylmercurial compounds in Iraq. It was a large epidemia, catastrophic in its size and number of lethal cases. Its outbreak in January 1972 was described by Bakir et al.[15] A total of 6530 patients were hospitalized, 500 of whom died. The population had consumed home-made bread, prepared from seeds treated with methylmercurial fungicides.

The distribution of the dressed seeds began in September 1971. In January 1972 several hundred poisoned patients were admitted to the hospital. From February up to the end of May, there were more than 5 cases daily. It is supposed that the exposure started with bread consumed since October. In January 1972, the authorities took strong measures to discontinue further consumption of treated seeds.

The mean methylmercury content in the flour was 9.2 µg/g, with variation from 4.8 to 14.6 µg/g. The poisoned subjects took in between 50 and 400 mg mercury, injested as methylmercury.

Mercury concentrations in the first blood samples were from 1100 to 3700 ng/cm³. The half-life of the agent was 65 d in the average and varied from 40 to 105 d.

A mercury content of 500 to 700 µg/g was found in the hair of patients. This method determined the beginning of exposure, depending on the length of the hair and its mercury content.

The clinical manifestations of the poisoning were characterized by the following main symptoms: ataxia, tremor, dysmetria, lack of coordination, emotional breakdown, visual impairments up to blindness, dysarthria, and auditory disturbances.

A connection was established between the symptoms and the quantities of ingested mercury. For instance, the lowest clinically manifested limit value for mercury was 25 mg, which was distinguished by the presence of paraesthesia. The limit values for ataxia, dysarthria, deafness, and lethal outcome were 55, 90, 170, and 200 mg, respectively (approximate mercury doses received by people during a several month consumption of contaminated bread).

The delayed appearance of symptoms, known for intoxications with methylmercury, had created a false feeling of security in the farmers. For instance, some had given treated seeds to chickens and did not note any immediate adverse effects. People who consumed flour from treated seeds, showed no symptoms of poisoning for weeks and months after consumption. The symptoms appeared only after mercury accumulated and the toxic dose was attained. It should be noted that the treated corn was indicated with red paint, which was easily washable. This misled the farmers into considering the usual washing sufficient.

The results of a great number of investigations on mercury in different media made it perfectly clear that the best indicator for its accumulation in human organisms is its determination in blood.

Methylmercury is excreted in milk in concentrations representing 3% of those determined in blood. These quantities may lead to damage in the breast-fed child if the concentrations in the blood of the mother are high. The newborn infants also can be injured in their embryonic development.[16]

Another case of mass intoxication with seeds treated with mercurial fungicides is described in Ghana.[17] Maize seeds were treated with methoxymethylmercury acetate as fungicide. It contained 2% ethylmercurychloride and 12% lindane. This fungicide was mixed with maize in a ratio of 1.8 g/kg. The treated corn had been carefully washed with hot water aimed at eliminating the pesticide. Then it was boiled and consumed in this form or as flour.

The typical cases of intoxication began with gastric colics, diarrhea, nausea, and vomiting. The latent period was from 5 to 15 d after the first consumption. All the patients had symptoms of disturbed CNS activity, formication of the extremities, weakness, tremor of hands and legs, insecure gait, and hyperreflexia. Children were more severely affected. Some of them had speech disturbances, others had paralysis of upper and lower extremities. A lethal outcome occurred as a result of dehydration, shock, and renal insufficiency. It was proved that the mercury content in urine did not correlate with the manifestations of poisoning and the lower mercury limit of symptom appearance could not be determined. The total number of the victims was 144, 20 of whom died.

Baybourina investigated children who had consumed some quantity of corn, treated with granosan.[18] In two thirds of the children toxic encephalitis and encephalomeningitis, with overt alterations in the vegetative nervous system, were diagnosed. At a repeated observation 6 months to 2 years later, in a part of the children, expressed hypothalamic symptomatics was found. It was manifested in vegetative-vascular form of diencephalon syndrome: deviations in the water and salt metabolism of diabetes incipidus type with thirst and dryness in the mouth, decreased heart rate, abundant perspiration, raised temperature, decreased 17-ketosteroids, nonmotivated fear, headache, emotional instability, deteriorated memory, etc.

Mouhtarova has studied 57 cases of chronic intoxication with granosan in children: 45 had consumed bread prepared from pesticide-treated grain; 2 were sucklings whose mothers had been poisoned by the agent, and the 10 others were from mothers had worked with granosan when pregnant.[19,20] The last group of children were born with gross cerebral changes and deficiencies in their mental and physical development; 6 of these cases were lethal.

The clinical picture of these intoxications enabled the author to draw the conclusion that chronic intoxication was more severe in children than adults. This is due to some developmental anatomic and physiological particularities of the organism and nervous system.

This study also revealed damage of the central and peripheral nervous system. Toxic encephalitis was diagnosed in 38 subjects, encephalomyelitis — in 4, and encephalopolyneuritis — in 5. Mainly the diencephalon-hypothalamic area and spinal cord were affected; the subcortical formations were affected less. The disturbances in hydrocarbon metabolism, found by Mouhtarova, were related to the diencephalon syndrome.[21]

Ramanauskayte observed neuroendocrine disturbances in 18 out of 67 cases of chronic intoxication of children: diabetes insipidus with mercury in urine from 0.002 to 0.03 mg/dm^3, adipose genital dystrophia, hypophyseal disturbances, and suprarenal insufficiency.[22] According to author data, at chronic intoxications by granosan, together with neurological and diencephalon phenomena, endocrine disturbances have been observed in 12% of the cases.

Sivitskaja described eye changes in 55 subjects who had been consuming meat and milk contaminated with methylmercury chloride during 3 months in the winter of 1968.[23] Their general condition was deteriorated. The patients had different nervous system damages, encephalo-polyneuritis, encephalopathia, asthenic syndrome, vegetative vascular dystonia, etc. Some of the patients complained of "flies" moving before their eyes. A decrease of visual acuteness was established, as well as dull cornea, hypertension, and sclerosis of the vessels of the retina.

Since early symptoms of intoxication with methylmercury are similar to those of polyneuropathy, electrophysiological methods investigate the peripheral nervous system for an early diagnosis. von Burg and Rustan[24] applied these methods in an investigation on 14 subjects, 5 months after being exposed to methylmercury; in the culmination of exposure, these subjects had had from 1000 to 3000 ng/cm^3 mercury in blood. At the moment of the investigation, mercury content in blood was 100 to 800 ng/cm^3. According to the authors, the clinical electrophysiological indices did not support the thesis that methylmercury intoxications produce effects similar to peripheral polyneuropathy. The methods of investigation they applied could not reveal any consistent identical disturbances in the different subjects affected. The only interesting data were those of the decreased threshold of "H" reflex. Some authors are of the opinion that the damage of the specific inhibitory centers in the low section of the brain stem could explain the easy stimulation of "H" reflexes.

VIII. TREATMENT

From the known mercury antidotes, D-penicillamine seems to be the most effective in intoxication with alkylmercury compounds.[15]

A polystyrene resin containing a fixed sulfhydryl group has been shown to reduce blood mercury levels in the intoxicated persons in Iraq.[15] Hemodialysis could help remove mercury from the organism. Antidote therapy and other measures to increase excretion of mercury should be ensured before irreversible damage has already occurred.

IX. PREVENTION

As far as the main use of organomercurial fungicides is for seed dressing, care should be taken to keep the equipment for treatment hermetic and use special vehicles for transportation.

Of special importance is decontaminating the containers and premises where treated seeds had been stored. This is necessary in view of the high absorption capacity of different materials for vapors. They have been found in concentrations from 2 to 6 mg/kg in the cement coat of storehouse rooms.[1]

The work clothing should also be subjected to special decontamination.[1]

X. CONCLUSION

Organomercurials are fungicides used mostly for seed dressing. They enter the body through the respiratory system and the skin during occupational exposure. The oral route is possible through intake of contaminated food.

This class of pesticides is very dangerous due to their high acute toxicity (LD_{50} from 30 to 80 mg/kg/b.w.), their strong potential for cumulation in human beings and through the trophic chain, and their polytropism of effect.

Their target organs and systems are the nervous system and the kidneys, but the cardiovascular system, liver, and hemopoiesis can also be affected. There is a risk for malformation if a pregnant woman is exposed. Organomercurials are allergens.

Exposure levels of up to 0.85 mg/m^3 were measured in working environment. Hair content in general population in some areas of Japan reached a mean of 23.4 mg/kg.

The excretion of mercury through human milk (0.04 mg/l at occupational exposure) may create health risks for the newborn.

Acute intoxications are rare and they are manifested by encephalopathy and gastrointestinal and renal disturbances.

Chronic intoxications are more common, mostly related to oral exposure. Clinical manifestations are mainly related to central and peripheral nervous system impairment: motor (ataxia, dysmetria, lack of coordination), emotional disturbances, sensory organ damage (blindness, hearing loss), and polyneuropathy. Specific antidote therapy is used (D-penicillamin, etc.) together with general symptomatic treatment.

The use of organomercurials in agriculture should be well controlled and limited.

REFERENCES

1. **Trakhtenberg, I. M. and Balashov, V. E.,** Organomercurials, in *Pesticides Manual,* Medved, L. L., Ed., Kiev, 1974, (in Russian).
2. **Vergieva, T.,** Organomercurial compounds, in *Manual for Safe Use of Pesticides,* Kaloyanova, F., Ed., Med. i Fizkult., Sofia, 1984, (in Bulgarian).
3. **Ueda, K.,** Discussion, *Whither Rural Medicine,* Proc. 4th Int. Contr. Rural Med., Usuda, 1969, Tokyo, 1970, 87.
4. **Wakatsuki, T.,** The acute state of pesticide poisoning in farmers in Japan, in 3rd Int. Congr. Rural Med., X, 1966, Bratislava, II, 79.
5. **Kasai, A.,** Studies in the analysis of pesticide residues in human body and foods, *Whither Rural Medicine,* Proc. 4th Int. Congr. Rural Med., Usuda, 1969, Tokyo, 1970, 86.
6. **Juszkievicz, T. T., Sprengier, T., and Radomeski, T.,** Stezanie rtecu w mleku kobiet, *Pol. Tyg. Lek.,* 30, 9, 365, 1975.
7. **Juszkievicz, T. and Sprengier, T.,** Pozostalosci rteci w nerkach koni joko wskaznik skazenia rtecia stodowiska, *Pol. Arch. Weter.,* 17, 1, 74, 1974.
8. **Sprengier, T.,** Stezania rteci w tkankach kotow z regionow rpzemyslawich rolniczych krajn, *Bromat. Chem. Toksykol.,* 7, 3, 371, 1974.
9. **Medved, L., Belonzhko, G., Mizjukova, J., Mukhtarova, N., Petrunjkin, V., and Trakhtenberg, J.,** Data on pathology and treatment of organomercury poisoning, in *Whither Rural Medicine,* Proc 4th Int. Congr. Rural Med., Usuda, 1969, Tokyo, 1970, 23.

10. **Bidstrup, P.,** *Toxicology of Mercury and Its Compounds,* Elsevier, Amsterdam, 1964, 250.
11. **Katzunuma, H., Suzuki, T., Nishi, I., and Kashima, T.,** Four cases of mercury poisoning, Rep. Institute for Science of Labour, 1963, 61.
12. **Patty, F.,** Industrial Hygiene and Toxicology. II. *Toxicology,* Interscience, New York, 1962.
13. **Mnatsakanov, T., Mamikonyan, R., Gevorkyan, P., and Nazaretyan, K.,** On electrolyte metabolism and ECG changes in chronic intoxications with granosan, *Gig. Tr. Prof. Zabol.,* 12, 7, 39, 1968, (in Russian).
14. **Taylor, W., Gurgis, H., and Stewart, W.,** Investigation of a population exposed to organomercurial seed dressing, *Arch. Environ. Health.,* 19, 4, 505, 1969.
15. **Bakir, F., Damluji, S., Amin-Zaki, L., Murtadha, M., Khalidi, A., Al-Rawi, N., Tikriti, S., Dhahir, H., Clarkson, T., Smith, J., and Doherty, R.,** Methylmercury poisoning in Iraq, *Science,* 184, 230, 1973.
16. **Gale, T. and Ferm, V.,** Embryopathic effects of mercuric salts, *Life Sci.,* 10, 1341, 1971.
17. **Derban, K.,** Outbreak of food poisoning due to alkyl-mercury fungicide, Food poisoning, *Arch. Environ. Health,* 28, 1, 49, 1974.
18. **Bayburina, S.,** On the question of the function of adrenals in children with hypothalamic damage of granozan etiology, *Gig. Primen. Toksikol. Pestits. Klin. Otravlenii,* 32, 1966, (in Russian).
19. **Mouhtarova, N.,** Some peculiarities of the toxic effect of granosan, *Gig. Primen. Toksikol. Pestits. Klin. Otravlenii,* 1965, 556, (in Russian).
20. **Mouhtarova, N.,** State of the treatment of nervous system in children at chronic intoxication with granozan, *Gig. Primen. Toksikol. Pestits. Klin. Otravlenii,* 431, 1967, (in Russian).
21. **Mouhtarova, N.,** State of some metabolic processes at chronic intoxications of nervous system with granozan, *Gig. Primen. Toksikol. Pestits. Klin. Otravlenii,* 334, 1965, (in Russian).
22. **Ramanauskajte, M. R. and Pikalitek,** On the question of the neuroendocrine disturbances at intoxications with organomercurial preparations, *Gig. Primen. Toksikol. Pestits. Klin. Otravlenii,* 502, 1966, (in Russian).
23. **Sivitskaya, I. and Provalova, A.,** The state of the organ of vision with sequelae of foodstuff intoxications by ethylmercurial chloride, *Gig. Primen. Toksikol. Pestits. Klin. Otravlenii,* 450, 1973, (in Russian).
24. **von Burg, R. and Rustam, H.,** Conduction velocities in methylmercury poisoned patients, *Bull. Environ. Contam. Toxicol.,* 12, 1, 81, 1974.
25. **Anon.,** Mercury Environ. Health Criteria 24, World Health Organization, Geneva, 1976.

DITHIOCARBAMATE COMPOUNDS

I. INTRODUCTION

Dithiocarbamate fungicides have found a large application. In general, they have low persistency in the environment, low toxicity, and no cumulative effects. Some of their metabolites, however, can create problems in the environment. The most serious among them is ethylenethiourea, which has high persistence and is potentially carcinogenic. A serious limitation of some dithiocarbamates is their teratogenicity for experimental animals.

II. PROPERTIES

Some metal salts and esters of methyl-, dimethyl-, ethylene-bis-dithiocarbamic acid are used as pesticides. Some properties of dithiocarbamate pesticides are shown in Table 1.

III. USES

Dithiocarbamates are used as contact fungicides for seed dressing (210 to 500 mg/kg) and plant treatment (0.25 to 0.5%, 2 to 10 kg/ha).

Some of them, e.g., vapam, are used as fungicides, nematocides and herbicides, and soil sterilants.

IV. METABOLISM

Dithiocarbamates enter the body mainly through the respiratory tract. They are metabolized to a number of toxic substances which are, to some extent, responsible for the intoxication, namely: carbon disulfide, hydrogen sulfide, dimethylamine and methyl isothiocyanate, ethylene thiourea, etc.

V. TOXICITY: MECHANISM OF ACTION

Toxicity of dithiocarbamates depends on the structure of individual compounds. Examples of low toxicity are zineb, LD_{50} 9 to 10 g/kg; maneb, 8 g/kg; ferbam, 17 g/kg; propineb, 5 g/kg; ziram, 1.4 g/kg; and vapam, 1.7 g/kg.[1,2] Nabam is one of the most toxic with LD_{50} 395 mg/kg.

One of the principal mechanisms of action of dithiocarbamates is the inhibition of enzymes containing SH-groups (alpha-ketoglutaroxidase, pyruvatdehydrogenase, succindehydrogenase, etc.) and the blockage of metal-containing enzyme systems due to helating of metal ions.

Dithiocarbamates and their metabolites have tyreostatic and teratogenic effects. Most of them are allergens.

VI. EXPOSURE: DOSE-EFFECT RELATIONSHIP

As dithiocarbamates are used mainly as powder formulations, occupational exposures could be very high.

For the general population, the main sources are food residues. Ethylenethiourea and other metabolites are found together with the parent compound.

TABLE 1
Properties of Some Dithiocarbamate Fungicides

Structural formula	Names and properties

Structural formula

$$H_2CNHCS \diagdown_{Zn} \diagup H_2CNHCS$$ (with S double bonds top and bottom)

Zineb
White crystal, m.w. 275.8, m.p. 240°C, practically insuluble in water and organic solvents, soluble in pyridin

Maneb
Yellow or brown powder, m.w. 265.3, m.p. 120°C, moderately soluble in water

Ziram
White crystal, m.w. 305.8, m.p. 246°C, soluble to 65 ppm in distilled water

Nabam
Colorless, m.w. 256.4

Ferbam
Dark brown powder, m.w. 416.5, m.p. 180°C, slightly soluble in water

Vapam (metam sodium)
White crystal powder, m.w. 130, solubility in water — 722 g/l at 20°C

TTD (tetramethyl thiuramdisulfide)
Grey-yellow powder, m.w. 240, m.p. 156°C, solubility in water 30 ppm

VII. EFFECTS ON HUMANS

Data concerning dithiocarbamate effects on man are scarce.

Skin irritation and sensibilization have been reported. Dermopathies have been described in persons engaged in zineb production and application. Bertolini observed the appearance of skin complaints after an 8- to 9-d contact with the formulation.[3] Skin disturbances are more frequent in subjects of advanced age and lowered resistence due to some past diseases (liver disease, avitaminosis, etc.). Skin lesions are enhanced by sunlight and wind, microtraumas by itching and the mechanical rubbing of clothes.

It is accepted that only a small proportion of dermatitis cases are of a contact type, most of them being predominantly allergic. Epicutaneous tests with zineb powder are strongly positive in subjects with previous exposure to the chemical, and negative in subjects without previous exposure. The experiments for developing passive immunity in guinea pigs and the sensibilization test of Drize have shown negative results.[3]

Bilanca et al.[4] reported cases of agricultural workers with dermopathies provoked by contact with zineb. Dermatologically, diffuse erythrodermia with erythema and the appearance of eczematoidic epidermitis of the eyelids, inguinal folds, and sexual organs were observed.

After recovery, sensibility toward sunlight persisted in the patients. Epicutaneous tests with 1% zineb solution were positive, confirming the close connection between the reaction and the formulation. After exposure to dithiocarbamates inflamed respiratory ways are found frequently.[5]

Pinkhas et al. reported cases of sulfhemoglobulinemia and acute hemolytic anemia with Hainz bodies in patients with glucose-6-phosphatedehydrogenase deficit and hypocatalasemia after contact with zineb.[6]

Tobacco workers had acute and chronic allergic skin lesions of the upper extremities due to contact with maneb-zineb treated tobacco plants. It was proved that only a part of the cases resulted from a directly provoked fungicide allergy. In most of the cases, the contact only triggered an already existing allergic trend in the organism.[7,8]

Orlov et al. investigated a group of workers exposed to ziram (57% were employed for more than 3 years).[9] Of the studied workers, 30% were proved to have disturbances in protein, hydrocarbon, and detoxication functions of the liver. This was manifested by a decreased total proteins, changed proteinogram (increase of β- and α-globulins), and hyperglycemia. In 8% of the subjects, inhibited secretory liver function with decreased uropepsin was observed, as well as radiologically proved hypertrophic gastritis. Dermatitis and eczema were diagnosed in 16% of the subjects, asthenovegetative syndrome — in 12.3%, rhinosinusitis — in 12.7%, and chronic bronchitis — in 5%.

Kaskevic described a clinical picture of thiuram (TMTD) intoxication, Jurchenko — neurological changes, and Larionov — blood morphology.[10-12] Persons occupationally exposed to thiuram in concentrations over TLV were investigated. The following complaints were registered: lacrimation, sneezing, rhinopharyngeal irritation, and headache. Less frequently, pains in the heart area, palpitations, vomiting, higher fatigability, nasal hemorrhage, and skin eruptions on the hands were observed, as well as itching on unclothed parts of the body. An increased alcohol sensibility with hyperemia of the face and upper part of the body, tachycardia, sweating, headache, nausea, and even collapse were also observed.

Cardiovascular symptoms, such as tachycardia, hypotonia, and myocardial dystrophy were observed. The dystrophic changes could result from metabolic myocardial injury on the one hand, and from increased serotonin concentration leading to coronary disturbances on the other hand. The involvement of a coronary factor also was confirmed by muscle hypoxy, observed in 25.5% of the subjects.

Functional disturbances of the CNS were found. Lung changes were diagnosed in 6.3% of the subjects; liver pathology, manifested by hepatocholecystitis and hepatitis of toxico-chemical etiology, was established in 33%. Gastritis with a secretory deficit was frequently found. Hyperplasia or dysfunction of the thyroid gland was found in 7.1% of the persons investigated. The increased reticulocyte number in peripheral blood indicated tension and a good regeneration ability of the bone marrow. Leukopenia, due mainly to a decrease of neutrophils and eosinophils was also present. Mihail et al. observed similar effects along with qualitative changes in leukocytes, e.g., fatty degeneration in the cytoplasm and barely noticeable neutrophilic granularity.[13] Anisocytosis and polysegmented neutrophiles with degenerative changes in nuclear structure were established in 78% of the subjects. In all cases significant fragmentation of the different neutrophiles segments was noted. Analogous changes were observed in lymphocytes.

Korablyov et al. reported leukopenic activity of dithiocarbamates.[14]

A severe intoxication with thiram has been described by Krupa.[16] Lesions and an inflammatory reaction of mouth mucosis, as well as gastromegalia and cholinesterase inhibition, were characteristic.

Clinically acute dithiocarbamate intoxications manifest themselves by headache, weakness, increased excitability, nausea, vomiting, diarrhea, and abdominal pains. In more severe cases, dizziness, depression, paralysis, and loss of consciousness are observed.

The treatment of intoxication is symptomatic. Inhalation with alkali solution helps the local bronchial irritation. Against systemic effects, vitamin C, glutamic acid, Ca gluconate, antihistaminic, and antiallergic treatment are recommended.

VIII. PREVENTION

Permissible values for some dithiocarbamates are as follows:[17,18]

TTD	0.5	mg/m³	MAC USSR
	5	mg/m³	TWA USA OSHA
	10	mg/m³	STEL USA ACGIH
	1500	mg/m³	IDLH USA
Ferbam	15	mg/m³	USA OSHA
	10	mg/m³	USA ACGIH
Zineb	0.5	mg/m³	MAC USSR
Maneb	0.5	mg/m³	MAC USSR

Acceptable daily intakes recommended by FAO/WHO are:[19]

Maneb	0.05	mg/kg/b.w./d
Zineb	0.05	mg/kg/b.w./d
Thiuram	0.005	mg/kg/b.w./d (temporary)
Ziram	0.02	mg/kg/b.w./d

During the first hours and days after application of dithiocarbamates it is possible to find high air concentration of CS_2, H_2S, and methyl isocyanate. Reentry in the treated fields should be done with caution and at least after a 6-d reentry period.

IX. CONCLUSION

Dithiocarbamates are a large group of contact fungicides. Almost all of them have low toxicity. Dithiocarbamates inhibit SH-groups containing enzymes and act as helating agents for metal-containing enzyme systems.

Human exposures do not represent serious health problems. Contact irritative, acute, and chronic allergic dermatites are often produced during occupational exposure. Liver function disturbances and cardiovascular and gastrointestinal system impairments are reported. Some deviations in hemopoiesis are also induced by dithiocarbamate exposure. Hyperplasia and thyroid gland dysfunction is a specific effect of this group of compounds.

REFERENCES

1. **Medved, A. I.,** Ed., *Handbook on Pesticides,* Urazhaj, Kiev, 1974, (in Russian).
2. *Farm Chemical Handbook,* 72nd Ed., Meister Publishing Co., Willoughby, 1986.
3. **Bartolini, E.,** L'action du zineb (ethylenbisdithiocarbamate de zinc) sur l'homme, *Med. Lav.,* 53, 1, 45, 1962.
4. **Bilanca, A.,** Erhythrodermia by zinc dithiocarbamate in an agricultural worker, *Arch. Ital. Dermatol. Venered. Sessuol.,* 33, 33, 1964.
5. **Hayes, W.,** Clinical Handbook on Economic Poisons, U.S. Department of Health, Education and Welfare, U.S. Government Printing Office, Washington, D.C., 1963.
6. **Pinkhas, J., Djaldetti, M., Joshua, H., and Resnik, C.,** Sulfhemoglobinémie et anémie hémolytique aigue avec corps du Heinz à la suite d'un contact avec un fongicide-zinc éthylène bisdithiocarbamate chez un sujet avec déficience de la glucose-6-phosphate déhydrogénase et hypothalassémie, *Blood,* 4, 484, 1963.
7. **Laborie, F. and Dedieu, R.,** Allergie aux fongicides de la gamme de manèbe et du zinèbe, Prophylaxie, enquête préliminaire, *Arch. Mal. Prof.,* 25, 78, 417, 1964.
8. **Laborie, F. and Laborie, R.,** Revue de pathologie comparée, Nouvelle série, La protection de la vie et de la santé, 66, 3-2-775, 105, 1966.
9. **Orlov, N., Kachlaj, D., Bayda, N., Djogan, I., Shtidlovskaja, V., Zinchenko, D., and Venetskaja, V.,** The effect of the derivatives of dithiocarbamic acid upon human organism, *Gig. Primen. Toksikol. Pestits. Klin. Otravlenii.,* 339, 1973, (in Russian).
10. **Kaskevic, L.,** Pathology of internal organs in persons working with TMTD, *Gig. Primen. Toksikol. Pestits. Klin. Otravlenii.,* 432, 1973, (in Russian).
11. **Yurchenko, I.,** Functional state of nervous system in workers engaged in seeds dressing, *Gig. Primen. Toksikol. Pestits. Klin. Otravlenii.,* 369, 1973.
12. **Larionov, V.,** Investigation of the morphological content in blood in subjects exposed to TMTD, *Gig. Primen. Toksikol. Pestits. Klin. Otravlenii.,* 388, 1973.
13. **Michail, G., Bodnar, J., Zlavoc, A., Branisteanu, D., and Ambrono, V.,** Researches concerning the exposure risk and the toxicology of one dithiocarbamic fungicide, *Whither Rural Medicine,* Proc. 4th Int. Congr. Rural Med., Usuda, 1969, (Tokyo), 1970, 63.
14. **Korablev, M., Vedenskij, V., and Manuha, L.,** Leukopenic activity of ethyl ether of dimethylthiocarbamate acid, *Zdravookhr. Beloruss.,* 10, 1964.
15. **Korablev, M.,** Tetramethylthiuramdisulphide — inhibitor of leukopiesis, *Toxicol. Pharmak. Pestic. Dr. Chim. Soed.,* 85, 1967, (in Russian).
16. **Krupa, A.,** Ciezkie zatrucie tiuramen, *Med. Wiejska,* 6, 1, 29, 1971.
17. **Burkackaja, E. N. and Zapko, V. G.,** Hygiene of work in application of toxic chemicals and mineral fertilizers in agriculture, Kiev, *Zdorov'e,* 1977. (in Russian)
18. The NIOSH Pocket Guide to Chemical Hazards, USD HNS, 1985.
19. **Vettorazzi, G.,** Ed., Pesticides Reference Index, JMPR, IARC-IPCS-IRPTC-VBC, 1981–1983.

Chapter 9

BENZIMIDAZOLE COMPOUNDS

I. INTRODUCTION

Pesticides from the group of benzimidazole are systemic fungicides. They have very low toxicity and good effectiveness in agriculture. Their stability in the environment and their long-term effect, such as mutagenicity and teratogenicity, produced by some of them in animals, impose some conditions and limitations in their use.

II. PROPERTIES

The chemical structure and some properties of benzimidazole fungicides are given in Table 1. They are easily degradable in the environment at alkaline pH, but they are stable in an acidic environment. Methyl-2 benzimidazole carbamate (carbendazim) is the principal metabolite of benomyl and thiophanate methyl in the environment.[1]

III. USES

Benzimidazole pesticides are systemic protective and eradicative fungicides; they are used as wettable powders and as a liquid formulation (40%) for microvolume air application. They are used for spraying strawberries, apples, cucumbers, vineyards, and tomatoes by 0.05 to 0.1% solution, 3 to 8 kg/ha. For air treatment of sugar beet, 0.6 to 1 dm^3/ha of preparation is used.[2]

IV. METABOLISM

No data on human metabolism are available. The main urine metabolite in animals for benomyl is methyl-N(5-oxibenzimidazole-2)carbamate, and for thiophanate—methyl-sulfate and glucoronide of 5-oxythiobendazole.

V. TOXICITY: MECHANISM OF ACTION

Acute toxicity of benzimidazole fungicides is very low. LD_{50} for carbendazin is 15,000 mg/kg, for benomyl — more than 10,000 mg/kg/b.w. for rats. Some benzimidazole fungicides are teratogenic and mutagenic for animals.

VI. EXPOSURE: DOSE-EFFECT RELATIONSHIP

Humans are exposed to benzimidazole derivatives via food.[2] In apples, 4 weeks after treatment, 0.1 mg/kg thiophanate methyl is found, in peaches, 100 d after treatment — 0.13 mg/kg and in grapes, 7 d after treatment — 4.13 mg/kg. Metabolites of thiophanate methyl are found in wine, as well.

Carbendazime is found in the milk and meat of cows fed with contaminated forage (10 and more mg/kg benomyl).

TABLE 1
Properties of Some Benzimidasol Compounds

Structural formula	Names and properties
	Carbendazim (Bavistin, Derosal, Equidazin) White solid Slightly soluble in water and oils
	Benomyl (Fundazol, Benlate) Mol. Mass 290.3, low vapor pressure, slightly soluble in water and oils
	Thiophanate methyl (methyl topsin) Mol. mass 342.4, slightly soluble in water and oils.

VII. EFFECTS ON HUMANS

The available information on human effects is related to the local and allergic reactions produced by benzimidazole compounds.

Benomyl is known to cause allergic contact dermatitis. Tomova[3] found 28% noncommunicable skin affections and 10.48% positive patch tests with 2% fundazole (benomyl) in 162 glasshouse workers. In the same number of control subjects, no positive test was found; however, in another group of glasshouse workers, positive reaction was also established in subjects exposed to other chemicals: unden (carbamate), zineb (dithiocarbamate), Decis (pyrethroid), and acrex (dinitrophenyl). Cross allergy reaction is likely to occur.

Later these findings were confirmed by the study of Matsushima and Aoyama.[4] In 41% of female farmers having used benomyl, positive reaction was found on patch testing (1% benomyl). Cross reaction between benomyl and diazinone, saturn, daconil, and Z-Bordeaut was demonstrated. In 20% of female workers who had not used benomyl, patch tests were positive. The theoretical basis for this phenomenon is unclear.

Dermatitis was experienced in 29 to 36% in the area where benomyl was used and in 23 to 25% in other areas.

Kühne et al. reported allergic dermatitis in female workers (7 from 40 exposed) after they had repeated contact with the benlate and chinoin-fundazol of a mushroom culture plant.[5] A follow up estimation 1 year later, showed a decreased response to patch tests. After 6 to 7 years, only two persons had strong sensitization, despite completely avoiding the contact with benomyl-containing formulations. Skin effects do not exist any more.

VIII. PREVENTION

For thiophanate methyl, MAC in U.S.S.R. is 1.5 mg/m^3, and for benomyl — 0.01 mg/m^3. ADI[6] for carbendazim is 0.01, for benomyl — 0.02 and for chlophanate methyl — 0.01 mg/kg/b.w./d.

IX. CONCLUSION

Benzimidazole compounds are newly developed systemic protective and eradicative fungicides with low acute toxicity. They do not represent hazard for acute intoxications. No data concerning chronic and long-term effects on humans are known. The available information on human effects is related mainly to allergic dermatitis.

Animal data on some long-term effects (mutagenicity and teratogenicity) should be considered in further follow-up studies, along with their application.

REFERENCES

1. *Farm Chemical Handbook,* 72nd Edition, Meister Publishing, Willoughby, 1986.
2. **Kaloyanova, F., Ed.,** *Handbook for Safe Use of Pesticides,* Med. i Fizkultura, Sofia, 1984.
3. **Tomova, L.,** Characteristics of skin sensitivity to the pesticide Fundazol, *Dermatol. Venerol.,* 18, 3, 164, 1979, (in Bulgarian).
4. **Matsushita, T. and Aoyama, K.,** Cross reaction between some pesticides and the fungicide benomyl in contact allergy, *Ind. Health,* 19, 77, 1981.
5. **Kühne, G., Hese, H., Plattke, B., and Puskabr, T.,** Dermatitis nach Benlate Kontakt, *Z. Gesamte Hyg.,* 31, 12, 10, 1985.
6. **Vettarazzi, G., Ed.,** Pesticides Reference Index, JUPR, IARC, IPCS, IRPTC-VBC, IPCS, 1961–1983.

Chapter 10

CHLORPHENOXY COMPOUNDS

I. INTRODUCTION

Herbicides from the group of chlorphenoxy compounds are relatively new substances. They were first marketed in 1948 and are widely used throughout the world. In recent years they have become very controversial, mostly due to the content of TCDD as a contaminant in Agent Orange, which was used in Vietnam. TCDD was found in 2,4,5-T, one of the components of Agent Orange, the other being 2,4-D. In experimental animals TCDD caused liver and kidney damage, chloracne, and long-term effects such as cancer and malformations.

Nowadays, improved technology is leading to decreased dioxins, especially 2,3,4,7,8-TCDD to the recommended WHO value of 0.1 ppm. In industry, more often the workers are exposed to mixtures of substances. In agriculture, or during the use of herbicides for other purposes, very often the exposure is mixed. These are the main limitations in evaluating the effects of chlorphenoxy compounds in humans.

II. PROPERTIES

A. CHEMICAL STRUCTURE

The group of chlorphenoxy compounds comprises a great variety of individual substances, as shown in Table 1. Usually, their alkaline (Na, K, NH_4) or amine salts and esters are used.

During production, if the process is poorly controlled, formation of polychlorinated dibenzo-p-dioxin by-products takes place. In some samples of 2,4-D, dioxins of about 20 to 4200 ppb are found, but not more than 1 ppb 2,3,7,8-TCDD. Meanwhile 2,4,5-T contains 2,3,7,8-TCDD, sometimes reaching up to 27 ppb. Current production specification limits TCDD to 0.1 ppm, following the recommendations of the WHO.

B. PHYSICAL PROPERTIES

2,4-D is a white powder with slightly phenolic odor. Some of the esters are liquids; 2,4,5-T is a white crystalline solid. Herbicide formulations are in the form of granules, dust, water soluble (alkali and amine) salts, and emulsifiable (esters) concentrates. Various solvents have been used with them. Some properties of 2,4-D and 2,4,5-T are summarized in Table 2.

III. USES

Chlorphenoxy compounds are used as agricultural herbicides to control broad leave weeds in cereals, green crops, roadside verges, and the area around farm buildings. Alkali and amine salts are used in about 0.5 to 1 kg/ha and esters at about 0.3 to 0.6 kg/ha.

At very low application rates (20 to 40 mg/l 2,4-D) they are efficient plant growth regulators on apple and citrus trees, potato plants, etc., usually when mixed with other substances[1-3].

IV. METABOLISM

Chlorphenoxyacetic acid compounds may be absorbed through the skin or inhaled during occupational exposure; in small amounts they may be absorbed orally. Dermal absorption is influenced by other substances present in the formulation, such as solvents, surfactants, etc.

The study of Libish et al. intended to determine the importance of the two exposure pathways for 2,4-D — the inhalation and dermal route.[2] Dermal absorption was found to be the major route,

TABLE 1
Chemical Structure of Chlorphenoxy Compounds

Structural formula	Chemical and common names
OCH_2 CO OH (ring with Cl, Cl)	2,4-dichlorphenoxyacetic acid 2,4-D; 2,4-PA; DCPA
OCH_2 CH_2 CH_2 CO OH (ring with Cl, Cl)	(2,4-dichlorphenoxy) butiric acid 2,4-DB
OCH, CO OH CH_3 (ring with Cl, Cl)	2(2,4-dichlorphenoxy) propionic acid dichlorprop; 2,4-DP
OCH_2 CO OH (ring with Cl, CH_3)	4-chloro-2-methyl phenoxyacetic acid MCPA - Dicotex
OCH_2 CH_2 CH_2 CO OH (ring with Cl, CH_3)	(4-chloro-2-methyl phenoxy) butiric acid MCPA
OCH, CO OH CH_3 (ring with Cl, CH_3)	α-(4-chloro-2-methyl phenoxy) propionic acid Mecoprop; MCPP
Cl — OCH$_2$ CO OH (ring with Cl, Cl)	2,4,5-trichlorphenoxyacetic acid 2,4,5-T
Cl — OCH, CO OH CH_3 (ring with Cl, Cl)	2-(2,4,5-trichlorphenoxy) propionic acid Fenoprop; 2,4,5-TP
Cl (dibenzodioxin ring with O, O, Cl, Cl, Cl, Cl)	2,3,7,8-tetrachlorodibenzo-p-dioxin TCDD

TABLE 2
Physical Properties of 2,4-D and 2,4,5-T

Property	2,4-D	2,4,5-T
Relative molecular mass	221	
Melting point	140–141°C	158°C
Solubility in water	Slightly soluble	Slightly soluble
Solubility in organic solvents	Soluble	Soluble
Vapor Pressure (Volatility)	52,3 Pa at 160°C	Low

up to 50 times greater than inhalation exposure with gun sprayers, and 4 to 11 times greater with mistblowers.

Workers using 2,4-D may absorb an average of 0.1 mg/kg/b.w./d.[3] 2,4-D is essentially unchanged by the body. Distribution in human tissues after heavy and fatal accidental intoxication is as follows:

Plasma	23–826	mg/l	Liver	100–293	mg/kg
Urine	146–420	mg/l	Lung	88	mg/kg
Brain	100–186	mg/kg	Heart	63–301	mg/kg
Kidney	315	mg/kg	Muscles	40	mg/kg

Values for the different tissues may vary according to elimination by treatment and time of death.

2,4-D and 2,4,5-T, excreted in urine, are unchanged. A small part is conjugated with the aminoacids glycin and aurin and glucoronic acid or is excreted as hydroxylated products or chlorphenol.[4,5] After oral administration of 5 mg/kg 2,4-D in male human subjects, 70% of the dose was excreted in the urine within 48 h.[3]

Excretion and complete elimination of 2,4-D may last about one week. For single occupational exposure the biological half-life of 2,4-D is about 25 to 48 h.

The clearance of 2,4,5-T from plasma and its excretion from the body, measured by its half-life, is about 1 d. The plasma contains 65% of the 2,4,5-T residue, where 98.7% are bound reversibly to protein.[1]

Applied to the forearm, only 5.8% of the ^{14}C labeled 2,4-D is excreted by the urine, but if applied intravenously, 100% is excreted within 13 h half-life.[7] The clearance rate for 2,4-D in humans is about 1 mg/kg/d.[6]

V. TOXICITY: MECHANISM OF ACTION

Chlorphenoxy compounds uncouple oxidative phosphorilation and decrease oxygen consumption in tissues. They disturb carbohydrate and other metabolism, as well as some endocrine functions (thyroid, suprarenals). They may inhibit certain ATP-dependent enzymes and thus affect lipid metabolism. 2,4-D decreases acetate metabolism; Chlorphenoxy compounds do not cumulate in the body and are moderately toxic. In long-term exposure, cumulation of the effect may damage the liver, kidney, and nervous system. They are irritants, and some are allergens.

VI. EXPOSURE: DOSE-EFFECT RELATIONSHIP

Occupational exposure in agriculture results in more than 0.1 mg 2,4-D kg/b.w./d.[3] Urine concentrations of chlorphenoxy compounds reflect total body burden, inhalation, and dermal and oral absorption. More available data indicated that 2,4-D and 2,4,5-T are essentially unchanged by the body and that urinary concentrations can be used as a guide for the degree of exposure.

TABLE 3
Exposure to Chlorphenoxy Compounds and Concentration in Blood and Urine

Exposure	Plasma	Urine	References
2,4-D occupational in forestry Exposure one week 0.1–0.2 mg/m³	—	8 (3–14) µg/ml or 9 mg total in 24 h urine	13
2,4,5-T occupational in forestry 0.1–0.2 mg/m³	—	4.5(1–11) µg/ml or 1 mg total in 24 h urine	
2,4-D occupational agricultural Exposure 0.002–0.01 mg/m³	—	1–20 mg/l	3
2,4-D oral, 5 mg/kg/b.w. (in milk or water)	10–30 mg/l 24 h Clearance t 0.5 = 10.6 h	Mean urinary excretion 0.5 = 17.7 h Total excretion: 82% of dose	3
2,4-D oral 5 mg/kg/b.w. in gelatin capsules	40 mg/l, 7–24 h Clearance t 0.5 = 33 h	75% excreted in 96 h	3
2,4-D oral 5 mg/kg/b.w.	35 µg/ml – 2 h 25 µg/ml – 24 h 3.5 µg/ml – 48 h	73% excreted for 48 h	6

Some data on the relationship between exposure concentration (in the air or orally) and plasma or urine concentrations are given in Table 3.

Occupational exposure differs in accordance with the type of equipment, personal protection, and type of formulation.

The exposure of council and forestry workers to 2,4,5-T has been monitored for 2 years by Simpson et al.[8] Inhalation exposure in relation to the type of spayer used was from 2 to 220 µg/h. Dermal exposure also varied in different parts of the body from 1 to 13,770 µg/100 cm³/h. The author accepted 100 µg/l 2,4,5-T in urine as an indication of excessive exposure, but not as an indication of poisoning. In this study he found from 160 to 1740 µg/l 2,4,5-T in urine during exposure and from 0 to 215 µg/l after cessation of exposure.

Lavy et al. studied 2,4,5-T exposure in the applicators and other crew members of helicopter and forest backpack spray teams, a tractor mist blower crew, and a tractor driver spraying rice levees.[9] Calculations based on gauze patch analysis indicated that external exposure values ranged from 0.0036 to 0.0136 mg/kg for helicopter crews, 0.081 to 1.85 mg/kg for the backpack crews, 0.046 to 0.205 mg/kg for mist blower crew, and 0.046 mg/kg for the tractor driver spraying the rice levees. Persons directly exposed under the spray potentially received 0.86 mg/kg.

In another study the ratio exposure/excreted 2,4,5-T in the urine was calculated to be 0.04. If this ratio is used, it could be assumed that 4% of the 2,4,5-T, which has been predicted to come in contact with an exposed dermal area, was actually absorbed and excreted in urine; the individual directly exposed to spray has absorbed and excreted 0.034 mg/kg 2,4,5-T. The authors calculated that the 2,4,5-T-quantity received by applicators should be below any possible health hazard levels.

Manninen et al. performed a field study during a tractor spray of MCPA (4-chloro-2-methylphenoxy acetic acid), dichlorprop (2-2-4-dichlorphenoxypropanoic acid) or dicamba (3,6-dichloromethoxybenzoic acid).[10] Air concentrations of phenoxy acid herbicides in the breathing zone were 0.027, within the range of 0.0011 to 0.2 mg/m³. Powder/water solution produced concentrations about 5 to 6 times higher than liquid preparation in water solution. Skin contamination, evaluated from patch samples ranged from 0.213 to 105 (mean 26.4) mg/h (outer aspects of clothes) and from 0.011 to 13.9 (mean 1.5) mg/h (inner aspect of clothes). Expiratory system exposure seemed to be lower than 13% of the total exposure. The urinary excretion of all herbicides had its peak 12 h after the exposure and was 5.73 mol/l (0.324 to 28.5 mol/l) or

TABLE 4
Exposure and Excretion of Herbicides

Herbicide	Air sample μg/m³ mean values (different equipment)	Urine sample mg/kg/urine	Amounts of herbicide inhaled and excreted		
			Mean amount (μg)		
			Inhaled	Excreted	%
2,4-D	7.1–55.2	1.64–1.77	44–331	1990–3570	2.2–9.3
Dichlorprop	9.2–17.2	1.54–6.14	55–103	1780–5140	2–3.1
Picloram	1.3–14.9	0.05–0.15	8–89	196–378	4.0–23.5

1.8 mg in a day. No significant difference was found in people exposed to powder and liquid formulation solution, which can be explained by the poor skin absorption of powder/water suspension. The increased excretion of sodium and potassium ions indicated the influence of the sodium-potassium pump in the kidney.

The biological half-life of MCPA is dose-dependent. It was about 12 h when the concentration of MCPA was more than 5 μmol/g creatinine and 30 h when the concentration was lower than 3 μmol/g creatinine.

The importance of the dermal exposure was underlined by Draper and Street, as well.[11] They studied 2,4,5-T and dicomba exposure in a ground boom spray application. The time weighted average for airborn herbicide residues did not exceed 2.2 μg/m³ in the cabs of vehicles. Dislodgeable dermal residues on the hand averaged 2.5 mg of dicomba and 7.5 mg 2,4-D.

Urine analysis showed that the maximum elimination of herbicides occured between 16 and 48 h after termination of exposure, but it continued even after 96 h. Urine concentrations of 2,4-D were to 20 mg/l, for dicomba — 0.5 to 16 mg/l, and for dicomba isomer — 0.39 to 15 mg/l. The dislodgeable dermal residues were of the same magnitude as these eliminated in the urine, which emphasizes the importance of the dermal exposure.

The report of Libich et al. deserves special attention.[2] He performed a quantitative assessment of inhalation, dermal absorption, and urine concentrations of different herbicides. Table 4 summarizes some of these data.

VII. EFFECTS ON HUMANS

A. ACUTE INHALATION AND DERMAL EXPOSURE

The chlorphenoxy compounds used as herbicides do not create serious health problems for workers during their production and use. They are moderately toxic, LD_{50} for different compounds being about 560 to 2900 mg/kg/b.w. As herbicides, almost all are skin irritants, and usually symptoms of irritation are a signal for high exposure leading to prevention from heavy intoxications.

Only poor occupational practice makes possible massive dermal and inhalation overexposure with signs and symptoms of acute intoxication.

Workers involved in building cleanup and equipment repair in a 2,4,5-T producing plant complained of skin, eye, and respiratory tract irritation, headaches, dizziness, and nausea. After 1 or 2 weeks, acneiform eruption, severe muscle pain affecting the extremities, thorax, and shoulders, fatigue, nervousness and irritability, dispnea, complaints of decreased libido, and intolerance to cold were registered.[13] Reviewing literature data, Paggiaro et al. stressed the following manifestations of 2,4-D acute intoxication by dermal and inhalation route: peripheral neuropathy, paraesthesia, pain in the extremities, hypoestesia to vibration and contact, hyporeflexia, reversible flacid paralysis, vomiting, diarrhea, mucohemorrhagia, high temperature, sweating, hyperazotemia, oliguria, proteinuria, erythema and edema of the mucoses.[14]

Paggiaro et al. reported a case of acute incidental poisoning by 2,4-D inhalation.[14] Its early manifestations were loss of consciousness with urinary incontinence, vomiting followed by myalgia, muscular hypertonia, headache, fever, and intermittent nodal tachycardia.

Skin damage was pronounced in some cases. A woman in Japan developed bullous contact dermatitis after 1 d occupational exposure to MCPA. The clinical picture progressed as a chemical burn with anuria and death occured 22 d later (after Bainova[15]). Lecombe et al. reported a case very indicative of dermal exposure.[16] A child playing in a garden treated with 2,4-D was intoxicated and the following symptoms appeared: vomiting, diarrhea, fever, flacid paraplegia, coma, dehydration, acetonuria, and azotemia. Similar are the cases of two girls playing in a garden treated with 2,4-D, 2,4,5-T and MCPA.[17] Irritation symptoms were predomiant: skin erythema, mucose membrane edema, oral mucosa inflammation, albuminuria, and oliguria.

B. ACUTE ORAL INTOXICATION

Acute oral poisoning by 2,4-D results in headache, dizziness, drowsiness, unsteadiness, dilated pupils, diplopia, vomiting, diarrhea, mucous burns, muscular weakness, impaired coordination and external stimuli response, speech difficulty, some muscle fibrilation, hyporeflexia, clonic spasms, convulsions, fever, cyanosis, heart disturbance, loss of consciousness, coma, and urinary incontinence. Laboratory analysis reveals leukocytosis, glycosuria, hyperglycemia, proteinuria, oliguria, hyperazotemia, and myoglobinemia. An autopsy finds degenerated convoluted kidney tubules, pulmonary emphysema and edema, liver necrosis, degenerative ganglion cells in the brain, widespread plaques of acute demyelinization in the brain, and myocardial injuries.[3,18-23]

Acute intoxication of 2,4,5-T results in weakness, lethargy, anorexia, diarrhea, muscle weakness, involving mastication and swallowing, ventricular fibrilation, and cardiac arrest.[1]

C. SHORT- AND LONG-TERM OCCUPATIONAL EXPOSURE
1. Irritative and Allergic Effects: Porphyria Cutanea Tarda

Respiratory tract irritation and dispnea with functional pulmonary disorders are common at occupational exposure. Bronchitis, peribronchitis, and pneumosclerosis have been connected with long-term occupational exposure.[24] Skin damage of different kinds have been reported: contact eczema and dermatitis, including anaphylactoid purpura (allergic angiitis). Suskind and Herbzberg found acneiform eruption (chloracne) in a longitudinal study of 204 workers from a 2,4,5-T production plant; it persisted for 30 years in 55.5% of them.[13] Poland et al. conducted health survey of 73 workers in a 2,4-D and 2,4,5-T plant; 7 employees complained of itchy eyes, 14 — frequent tears, 5 — bloodshot eyes, and 7 — sties.[25] A physical examination found 4.1% had appreciable conjunctival injection, 31.5% had hyperemia of the nasal mucosa, and 10.9% had inflammation of the buccal mucosa. Active acne was found in 13 (18%) of the workers. Hyperpigmentation (30 workers) and hirsutism (16 workers), most pronounced on the face, correlated with the severity of active acne and a high score on the manic scale of the Minnesota Multiphasic Personality Inventory.

Porphyria cutanea tarda was reported in previous studies of the same plant. This syndrome consists of liver dysfunction, vesiculobulous lesions in areas exposed to light, hirsutism, excessive mechanical skin fragility, hyperpigmentation, excretion of red urine containing an increased amount of uroporphyrin, coproporphyrin, or both, and a large increase in urinary deltaaminolevulinic acid (ALA) or porphobilinogen (PBG).[25]

In the study by Polland et al., no clinical porphyria could be documented, and only one worker had persistent uroporphyrinuria.[25] The excretions of coproporphyrin, ALA, and PBG, although not abnormally elevated, increased in the 15 maintenance men, relatively.

An epidemiological case-control study of forestry workers exposed to 2,4,5-T was performed by van Houdt et al.[26] Compared to a matched group of nonexposed workers, they showed no significant differences in acne, sores or urinary porphyrin patterns.

It is now accepted that chloracne and porphyria cutanea tarda are caused by the impurities in 2,4,5-T and 2,4-D, and especially by 2,3,7,8-TCDD.[1,3,6,13]

2. Gastrointestinal System

An epidemiological study of 241 workers in 2,4,5-T production plant reported the association between exposure and the history of gastrointestinal tract ulcer.[13] Polland et al. found functional disorders or subjective symptoms in 30% of the workers in their study, such as nausea, vomiting, diarrhea, abdominal pains, and blood in stool.[25]

Anorexia and gastric pains often have been found in occupationally exposed agricultural workers.[6] A higher morbidity by diseases of the digestive tract and liver has been reported in 2,4-D production.[6]

3. Nervous System

There are many reports on the toxic effects of chlorphenoxy compounds on the central and peripheral nervous system.[13,26-29] In some cases other coincidental causes are suspected to produce such effects; further studies are needed in this field.

The reported effects on the nervous system may be grouped as follows:

- Central nervous system effects: headache, fatigue, drowsiness, impaired coordination, diminished proprioception, and electroencephalographic abnormalities (bilateral higher activity and theta rhythms paroxism)[24]
- Sensory neuropathy: decreased hearing, impaired taste sensitivity, hyposmia, intolerance to certain odors, and hypersensitivity to noise
- Peripheral neuropathy: hyporeflexia, muscular weakness, reduced peripheral nerve conduction velocity, and partial paralysis.

In the study of Polland et al., 27% of 73 examined workers complained of lower extremity fatigue or difficulty in climbing stairs.[25] Goldstein et al. reported peripheral neuropathy with paraesthesia and paralysis in three farmers after occupational exposure.[30] Recovery was incomplete even after a year. Electromyographic examinations supported the diagnosis of peripheral neuropathy.

4. Psychological Effects

Polland et al. performed a special study using the MMPI scale.[25] Positive results were obtained for hypochondria, depression, hysteria, psychodeviance, paranoia, and mania. Extensive fatigue, emotional instability, diminished learning ability, and a loss of vigor and libido were also reported.

5. Hepatotoxic Effects

Polland et al. found some liver tests positive in long-term exposed workers.[25] Of the 73 employees, 2 had alkaline phosphatase values slightly above the normal range, 2 others had elevated SGOT levels, and 1 had a palpable liver. Hyperbilirubinemia, increased urobilinogen levels, and an enlarged liver have been reported in workers occupationally exposed to 2,4-D and other chemicals.[13,28,29] Inhibited mixed function oxidase activity was reported by Krasnyuk and Lubjanova.[24]

6. Myotoxic Effects

Myotonia, fibrilation, stiffness, muscle damage, and atrophy have been reported in 2,4-D intoxications.[27,28,31] Although the case reported by Berwick was an accidental oral intoxication with 30 ml 2,4-D mixed with other substances, the symptoms described by the author are relevant for other routes of absorption.[31] According to the author fibrilarity, twitching and transient paralysis of intercostal muscles, and generalized skeletal muscle damage were demonstrated by dramatically elevated SGOT, SGPT, LDH, aldolase, and creatinophosphoki-nase levels (36, 20, 24, 49 and 69 times, respectively, compared with normal values). Creatinine phosphokinase is a catalytic enzyme mediator for creatinine, adenosine triphosphate, and adenosine diphosphate; its highest concentration is normally found in skeletal and heart muscle. It appears in very high concentration in the serum of patients with muscular diseases. The appearance of oxymyoglobin in the urine supported the hypothesis of Berwick. Myogloburin-uria and hemoglobinuria are to be tested as indicators of myotoxic effect.

7. Cardiopathy and Cardiovascular Effects

It is not clear whether the effects on the cardiovascular system are connected with myocardial impairment, nerve conductance or increased sensitivity due to allergic reaction. Functional changes, myocardial dystrophy, myocarditis, cardiac arrhythmia, ECG changes, hypotonia and hypertonia, have been reported after occupational exposure.[3,14,24,29]

8. Endocrine Effects

Workers packing 2,4-D sodium salts showed impaired iodine uptake by the thyroid and decreased thyroxine, thyroxine clearance, and thyroxine iodine values.[3]

9. Hematological and Biochemical Effects

Moderate anemia with changes in erythrocyte size or volume, methemoglobinemia, monocy-tosis, lymphocytosis, eosinophyllia, hypo- or hyperglycemia, hypercholesterolemia, elevated phospholipids, altered blood albumin and globulins, elevated blood urea level, transaminase (SGOT, SGPT) creatinine phosphokinases, delayed prothrombin time, and decreased SH-groups are reported.[3,13,24,28]

10. Long-Term Effects

The reproductive effects found in animals (birth defects and reduced testosterone uptake by the prostate) were not confirmed by epidemiological studies on workers exposed to 2,4-D or 2,4,5-T.

Constable and Hatch reviewed the unpublished Vietnamese epidemiological studies on the reproductive effects of 2,4,5-T (Agent Orange).[32] Two types of studies were carried out in Vietnam: (1) studies comparing the reproductive outcomes among couples living in sprayed areas (South) with those among couples living in unsprayed areas and (2) studies comparing the reproductive outcomes among women and men living in the North and exposed women whose husbands served in the South. The following outcomes have been examined: miscarriage, stillbirth, congenital defects, and hidatidiform mole or molar pregnancy.

Studies in North Vietnam demonstrated consistency in reporting an association between presumptive paternal exposure to herbicides prior to conception and congenital defects in subsequent offspring, particularly certain types of anomalies (anencephaly and orofacial defects). Studies in South Vietnam reported increases in miscarriages, stillbirths, and birth defects. Authors from other countries (USA and Australia) have reviewed the publications as well. They found no association between reported Vietnam service and adverse reproduction effects.

The authors concluded that the measurement of the exposure had been lacking or quite crude. The timing of the exposure in relation to pregnancy had not been well evaluated in the study.

Neither of the studies on mutagenic effects are conclusive. Coggon and Achelson concluded that literature data suggests a biological association between phenoxy herbicides and soft tissue sarcoma; weaker evidence relates these products to malignant lymphoma.[33] This evidence is based mostly on Swedish case control studies (110 soft tissue sarcoma).[34,35]

TABLE 5
Dose-Effect Relationship

Dose	Effect
2,4-D	
30 mg/kg/b.w./d (purified 2,4-D)	No ill effects
500 mg/man/d for 21 d (purified 2,4-D)	No ill effects
5 mg/kg/b.w. (volunteers)	No effect
300 mg/man (orally)	Acute toxic effect on digestive tract
3.6 g/man (orally)	Acute intoxication
6.5 g/man	Lethal effect
80 to 800 mg/kg/b.w.	Lethal effect
2,4,5-T	
10 mg/kg/b.w./d (2 years; animal experiment)	No effect level
3–4 g/man	Toxic effect

D. EXPOSURE/EFFECT RELATIONSHIP

The data on the relationship between occupational exposure to chlorphenoxy compounds and health effects are very limited. Table 5 summarizes the data available on dose effect relationships in humans.[1,3,6,18,22]

An epidemiological study of 220 men exposed to 2,4-D in a manufacturing plant did not reveal any adverse health effects; they were exposed from 0.5 to 22 years with an average daily intake of 30 to 40 mg.[6] Four men, investigated during their work week at 0.1 to 0.2 mg/m³ 2,4-D, showed no clinical symptoms.[36]

VIII. SAFE LEVELS

There is no guideline for permissible urine concentrations of chlorphenoxy compounds. Libich et al.[2] use urinary guides which reflect dose equivalent to that received by inhalation alone, assuming that concentration is equal to 10 mg/m³ TLV. Using an urinary model, the authors calculated that, during weekly exposure, Thursday or Friday will reflect the average daily exposure. Transforming urinary 2,4-D concentration to equivalent daily exposure to airborn 2,4-D is based on a standard man who inhales 10 m³ of air during an 8 h working period and eliminates 1.4 l of urine/d. The authors proposed the following urinary concentrations as a guide for exposure:

Grade of exposure	2,4-D mg/kg in urine
Normal	up to 0.7
Low	from 0.7 to 7
Moderate	from 7 to 70
High	more than 70

Low grade of exposure corresponds to 0.1 to 1 mg/m³ inhalatory exposure and high grade — more than 10 mg/m³.

The air concentration of 2,4-D and dichlorprop averaged only 0.1% TLV. According to the equipment used in these studies, urine excretion varied from 2 to 3.6 mg/d for 2,4-D and from 1.8 to 5.2 mg/d for dichlorprop. Compared to the proposed guide, the workers show low exposure.

Libich et al. gave a detailed method for urine analysis of 2,4-D.[2] They recommend sampling urine the afternoon of the day of exposure or the next day. When exposure lasts longer, urine levels in samples on the fourth or fifth day of exposure reflect the daily exposure better. Sensitivity of the method is 0.01 ppm.

Grover et al. described the procedure for determining 2,4-D and dicamba in inhalation, dermal, hand-wash, and urine samples.[37] Residues are derived with diazomethane and quantitated using electron capture detector gas chromatography. Recovery is 80% at the limit of detection. The limit of detection for 2,4-D in urine is 5.0 μg/100 ml, and for dicamba it is 1.7 μg/100 ml.

During the day of 2,4-D application in agriculture, urine levels are 8 μg/ml; they are 3 to 4 μg/ml the afternoon following a day of exposure. For 2,4,4-T, the urine levels are 4.5 μg/ml and 1 to 11 μg/ml, respectively. In another investigation, the mean excretion of urine in 24 h was 9 mg 2,4-D and 1 mg 2,3,5-T. Plasma levels of both herbicides reached 0.1 to 0.2 μg/ml. No clinical symptoms were associated with these levels.[12,34]

Urine excretion reaching up to 5.1 mg/d dichlorprop and 3.6 mg/d 2,4-D does not provoke any symptoms.[2] A quantity of 335 mg 2,4-D/l plasma does not produce symptoms. The acute lethal level of 2,4-D in plasma is probably between 447 and 826 mg/l.

The following data should be kept in mind when discussing permissible urinary levels:

- 70 mg/kg urine 2,4-D corresponds to U.S. TLV/2,4-D.
- 7 mg/kg urine 2,4-D corresponds to U.S.S.R. MAC/2,4-D.
- 0.3 mg/kg/b.w. 2,4-D is the acceptable daily intake recommended by FAO/WHO IMPR.
- 0.1 mg/kg/b.w./d 2,4-D is the mean occupational exposure.
- 0.026 mg/kg/b.w./d 2,4-D is the mean general population exposure.
- 0.1 mg/l urine 2,4-D is accepted by Simpson et al. as an arbitrary standard.[8]

There are some limitations in the first two values, because they are calculated in the unrealistic condition that all amounts in the inspired air will be absorbed. The only conclusion from these data is that the urine level of 2,4-D should not be greater than 7 to 70 mg/kg urine. Epidemiological studies suggest that levels of 2,4-D in urine reaching about 10 mg/l are safe.

The following are recognized maximum permissible limits in various countries:

- USA — TLV (ACGIH) for 2,4-D, dichlorprop, and 2,4,5-T is 10 mg/m^3
- U.S.S.R. — for 2,4-D MAC is 1 mg/m^3, reentry period is 10 d; MAC for water is 0.01 mg/l
- WHO/FAO — recommended acceptable daily intake (ADI) for 2,4-D is 0.3 mg/kg/b.w./d

The immediate dangerous to life or health (IDLH) level for 2,4-D is 500 mg/m^3. This level represents a maximum concentration from which one could escape within 30 min without any impairment symptoms or irreversible health effects.[38]

IX. CONCLUSION

Chlorphenoxy compounds are widely used herbicides. The principal concern about their use is the presence of some dioxins as impurities, especially 2,3,4,7,8 TCDD. That is why the World Health Organization recommended the maximum allowable value of dioxin to be less than 0.1 ppm.

Chlorphenoxy compounds enter the body through inhalation and the skin during occupational exposure. Their mechanism of action is related to uncoupled oxidative phosphorylation and decreased oxygen consumption in tissues, as well as to disturbances in carbohydrate and other metabolic processes. They are moderately toxic and do not create serious health problems for workers.

Chlorphenoxy compounds may produce acute and chronic intoxications. Irritative and allergic effects and the development of porphyria cutanea tarda are typical of this group of compounds.

The gastrointestinal system and the central and peripheral nervous systems may also be affected. Peripheral neuropathy with paresthesia and reversible paralysis is reported. Myotic effects result from muscle damage and atrophy. Other systems and organs are also affected, such as the liver, cardiovascular system, thyroid gland, and hemopoietic system.

Studies on teratogenic and mutagenic effects are not conclusive.

REFERENCES

1. 2-4-5-T Data Sheet on Pesticides No. 13, VBC/DS/75, 13, Food and Agriculture Organization/World Health Organization.
2. **Libich, S., James, C. T., Frank, R., and Siron, G.,** Occupational exposure to herbicides used along electric power transmission line right of way, *Am. Ind. Hyg. Assoc. J.,* 45, 56, 1984.
3. **Environmental Health Criteria,** 2,4-Dichlorphenoxyacetic acid (2,4-D), Environmental Health Criteria No. 29, World Health Organization, Geneva, 1984, 151.
4. IARC Monograph on the evaluation of the carcinogenic risk of chemicals to human, 15, 1977.
5. WHO Pest. Res. Ser., No. 4, World Health Organization, Geneva, 1985.
6. 2,4-D Data Sheets on Pesticide No. 37, VBC/DS/78, 37 Food and Agriculture Organization/World Health Organization.
7. **Feldman, R. I. and Maibach, H. L.,** Percutaneous penetration of some pesticides and herbicides in man, *Toxicol. Appl. Pharmacol.,* 28, 126, 1974.
8. **Simpson, G. R., Higgins, V., Chapman, I., and Bermingham, S.,** Exposure of council and forestry workers to 2,4,5-T, *Med. J.,* August 2, 546, 1978.
9. **Lavy, T. L., Shepard, J. S., and Bouchard, D. C.,** Field worker exposure and helicopter spray pattern of 2,4,5-T, *Bull. Environ. Contam. Toxicol.,* 24, 1, 90, 1980.
10. **Manninen, A., Kangas, J., Klon, T., and Savoleinen, H.,** Exposure of Finnish farmworkers to phenoxy acid herbicides, *Arch. Environ. Contam. Toxicol.,* 15, 1, 107, 1986.
11. **Drapper, W. M. and Street, J. C.,** Applicator exposure to 2,4-D dicamba and dicamba isomer, *J. Environ. Sci. Health,* 17, 4, 321, 1982.
12. **Kolmodin-Hedman, B., Erne, K., and Akerblom, M.,** Field application of phenoxy acid herbicides, in *Field Workers Exposure During Pesticides Application. Studies Env. Sc. 7,* Turdoir, W. F. and Haemstra, E. A., Eds., Elsevier, 1980, 73.
13. **Suskind, R. R. and Hertzberg, V. S.,** Human Health Effect of 2,4,5-T and its toxic contaminants, *JAMA,* 251, 18, 2372, 1984.
14. **Paggiaro, P. L., Martino, E., and Mariotti, S.,** On a case of poisoning by 2,4-dichlorphenoxyacetic acid, *Med Lav.,* 65, 128, 1974.
15. **Baynova, A.,** Chlorphenoxy compounds, in *Manual of Safe Use of Pesticides,* Kaloyanova, F., Ed., Sofia, 148, 1983.
16. **Lecombe, A., Farrious, R. P., Fontaine, G., and Muller, P. H.,** A propos d'un nouveau cas d'intoxication par le 2,4-D, *Rev. Pediatr.,* 3, 4, 207, 1967.
17. National Clearinghouse for Prison Control Center. Deaths from chlorinated phenoxyacetic 2,4-D, 2,4,5-T, MCDA, U.S. Department of Health, Education and Welfare, March-April 1968.
18. **Dudley, R. A. and Thapar, N. T.,** Fatal human ingestion of 2,4-D — a common herbicide, *Arch. Pathol.,* 94, 270, 1972.
19. **Curray, A. S.,** Twenty one uncommon cases of poisoning, *Br. Med. J.,* 1, 687, 1962.
20. **Seabury, J. H.,** Toxicity of 2,4-dichlorphenoxyacetic acid for man and dog, *Arch. Environ. Health,* 7, 202, 1963.
21. **Jones, D. J. R., Knight, A. J., and Smith, A. J.,** Attempted suicide with herbicide MCDA, *Arch. Environ. Health,* 2, 14, 363, 1967.
22. **Nielsen, R., Raemse, B., and Holme, I. I.,** Fatal poisoning in man by 2,4-dichlorphenoxyacetic acid (2,4-D). Determination of the agent in forensic materials. *Acta Pharmac. Toxicol.,* 22, 224, 1965.
23. 2,4-D SC. Reviews of Soviet literature on toxicity and hazards of chemicals, No. 70. UNEP, IRPTC, USSR Commission for UNEP. Centre of International Projects GKNT, Moscow 1985.

24. **Krasnjuk, E. M. and Lubjanova, I. P.,** Occupational diseases in agricultural workers in *Zdorovje,* Kundiev, I. I. and Krasnjuk, I. P. Eds., Kiev, 1983, 63.
25. **Polland, A., Smith, D., Metter, G., and Passik, P.,** A health survey of workers in a 2,4-D, and 2,4,5-T plant with special attention to chloracne, porphiria cutanea tarda and psychologicial parameters, *Arch. Environ. Health,* 22, 316, 1971.
26. **Van Houdt, J. J., Fransman, Z. G., and Strik, J. J.,** Epidemiological case control study in personnel exposed to 2,4,5-T, T.W.A. "Chemisphere", 12, 575, 1983.
27. **Fetisov, M. I.,** Occupational hygiene in the application of herbicides of the 2,4-D group, *Hyg. Sanit.,* 31, 383, 1966.
28. **Bahirov, A.,** The health of workers involved in the production of amine and butyl-2-4D herbicide, *Vrach. Delo,* 10, 92, 1969.
29. **Zaharov, G. G., Shevchenko, M., Zub, G. A., and Pershal, L. K.,** The clinical picture of prophylaxis of Dicotex poisoning, *Zdrav. Beloruss.,* 9, 7, 1968.
30. **Golstein, N., Jones, P., and Brown, J.,** Peripheral neuropathy after exposure to one ester of dichlorphenoxy-acetic acid, *JAMA,* 130, 1307, 1959.
31. **Berwick, P.,** 2,4-Dichlorphenoxyacetic acid poisoning in man, *JAMA,* 214, 1114, 1970.
32. **Constable, J. D. and Hatch, M. C.,** Reproductive effect of herbicide exposure in Vietnam: recent studies by the Vietnamese and others, *Teratogen, Carcinogen, and Mutagen.,* 5, 231, 1985.
33. **Coggon, D. and Acheson, E. D.,** Do phenoxy herbicides cause cancer in man, *Lancet,* May, 8, 1982.
34. **Hardell, L. and Sandstrom, A.,** Case control study of soft tissue sarcoma and exposure to phenoxyacetic acids or chlorphenols, *Br. J. Canc.,* 39, 711, 1979.
35. **Erikson, M., Hardell, L., Berg, N., Möller, T., and Axelson, O.,** Soft tissue sarcomas and exposure to chemical substances: a case referent study, *Br. J. Med.,* 38, 27, 1981.
36. **Kolmodin-Heldman, B. and Erne, K.,** Estimation of occupational exposure to phenoxy acids, paper at the 21st Congr. of Eurp. Toxic. Society, Dresden, 1979.
37. **Grover, R., Cessne, A. I., and Kerp, L. A.,** Procedure for the determination of 2,4-D and dicamba in inhalation, dermal, hand-wash and urine samples from spray applicators, *J. Environ. Sci. Health,* 20, 1, 113, 1985.
38. **Anon.,** NIOSH Pocket Guide to Chemical Hazards, U.S. Department of Health and Human Services, September, 1985.

Chapter 11

DIPYRIDILIUMS

I. INTRODUCTION

Only two compounds from the group of dipyridiliums are used as pesticides — paraquat (since 1958) and diquat. Despite the irreversible lung effects produced by paraquat, these dipyridilium compounds are used in many countries (more than 130 for paraquat and more than 100 for diquat). The first reason they are used is their high biological activity as herbicides, which is similar to hormonal action. The other reason is their easy degradability in the environment. The products of their photodegradation are not toxic and do not migrate in plants. They are absorbed by soil particles and bind to clay materials to form biologically nonactive residues, which later are degraded by soil microorganisms or chemical reactions. Paraquat and diquat residues in plant and animal products are negligible. Paraquat and diquat are produced in the U.S., U.K., Italy, China, Japan, etc.

II. PROPERTIES

Chemically, paraquat is 1,1′-dimethyl-4,4′-bipyridilium dichloride. Its structural formula is given below:

$$CH_3-^+N\hspace{-0.5em}\bigcirc-\bigcirc\hspace{-0.5em}N^+-CH_3 \cdot 2Cl^-$$

Synonyms of paraquat are: Gramoxone, Dextront, Esgram, Katalon, Orvar, Seythe, Weedrite, and Weedol.

The specific gravity at 20°C is 1.24 to 1.26. The melting point is 175 to 180°C, and the boiling point is at 300°C with decomposition Paraquat has a very low vapor pressure. The solubility in water at 20°C is 700 g/l. Paraquat is insoluble in organic solvents and slightly soluble in alcohol. It readily undergoes a single electron reduction to a cation radical. Paraquat is formulated as a liquid (dark aqueous solution) — 20%, and as granules — 20 g/kg.

Diquat is 1,1′-ethylene-2,2′-bipyridilium dibromide. Its structural formula is given below:

$$\bigcirc\hspace{-0.5em}=N^+ \quad ^+N\hspace{-0.5em}=\bigcirc \quad 2Br^-$$
$$CH_2-CH_2$$

Synonyms of diquat are: reglone, aquacide, reglox, weedtrim, and deiquat. The specific gravity at 20°C is 1.2, the melting point is 180°C, the boiling point is about 300°C with decomposition, and the solubility in water at 20°C is 700 g/l; vapor pressure is very low.

Paraquat and diquat are stable in neutral and acid solutions, but they are hydrolyzed by alkali. Paraquat readily undergoes a single-electron reduction to a cation radical.[1]

III. USES

Paraquat and diquat are total contact herbicides. They kill broad-leaved and grassy weeds. Paraquat is used as a dessicant and defoliant on certain crops: potato, sunflower, sugar-cane, and cotton. At higher concentration it is effective in cleaning railways and roadsides weeds.[1] Diquat controls floating weeds in water bodies (lakes, drainage ditches, and irrigation ditches).

The application rate for paraquat and diquat ranges from 250 to 1500 g/ha, up to 2200 g/ha in special cases. The working solutions vary from 0.1 to 0.5%. They affect only the green parts of the plant due to photosynthesis blockage.[1]

IV. METABOLISM

Paraquat and diquat enter the human body through oral, pulmonary, and dermal routes. Dermal absorption is enhanced if the skin is damaged. The greatest part of ingested paraquat is eliminated in feces.

In the organism, paraquat is circulated to all organs and tissues. The storage of paraquat is higher in the lungs, which accumulate paraquat in plasma.

The kidneys eliminate paraquat from the blood and organs. This is one reason for the early onset of renal damage and, in severe cases, renal failure. As a result, removing paraquat from organisms should become very difficult and slow; this will facilitate its accumulation in the lungs.[1]

Paraquat does not undergo serious metabolic transformation, and it is excreted unchanged in the urine and bile. In the organism, it participates to a considerable extent in cyclic reduction-oxidation reactions. Due to the loss of one electron, a free radical of paraquat is formed; it is oxidized to the parent compound.

Diquat is excreted unchanged, and about 5% of the oral dose is eliminated as diquat monopyridone in feces and diquat dipyridone in urine.[1]

V. TOXICITY: MECHANISM OF ACTION

Data on animals demonstrate high toxicity of paraquat. Oral LD_{50} for guinea pigs is 30 mg/kg, for rats it is 100 mg/kg, and for rabbits LC_{50} (4 h) is 6.4 mg/m^3 of air. Rats exposed 6 h daily over 3 weeks showed lung irritation at 0.4 µg/m^3.[2]

For diquat, LD_{50} in rats is 231 mg/kg. A 6-h exposure of 1.06 µg/l air concentration for 1 to 5 d showed no adverse effect.[3]

The calculated oral LD_{50} for paraquat for men is 35 to 40 mg/kg.b.w. equivalent to 15 ml 20% solution or 25 ml 12% solution.[1,4]

The toxic action of paraquat is related to single electron reduction and oxidation reaction with chlorophyll, flaveproteins, and molecular oxygen.[5] It depletes (oxidizes) cellular NADPH and generates free radical and the superoxide ion (O_2^-), peroxidizes lipid cellular membranes, depolymerizes hyaluronic acid, and causes proteins and DNA inactivity.[1] NADPH depletion may disturb NADPH-dependent biochemical processes such as fatty acid synthesis.[1] The lungs are the most sensitive targets because they have the closest contact with atmospheric oxygen.

VI. EXPOSURE: DOSE-EFFECT RELATIONSHIP

Occupational exposure to paraquat and diquat is the most important. Different methods of applying paraquat are used; exposure of the people engaged in spraying has been evaluated.[6-10] Dermal pads consisted of polyethylene-backed 100 cm Whatman grade 542 filter papers attached to the skin or clothing.

Swan studied paraquat exposure during two spray operations in a Malaysian rubber plantation using hand-operated knapsack sprayers.[6] Urinary paraquat concentrations over a 12 week period were used as a biological indicator of exposure. The highest concentration was 0.32 ppm (mg/l), and most of the urine contained less than 1 ppm. In the second trial a total of 394 samples were analyzed. Paraquat was detected in only 53, and the highest concentration was 0.15 ppm.

Later knapsack spray operators, carriers, and rubber tappers in a Malaysian rubber plantation were studied for occupational exposure to paraquat.[7] The mean unclothed exposure was 2.2 mg/h (0 to 12.6 mg/h) and the overall total exposure was 66 mg/h (12.1 to 169.8 mg/h). The mean paraquat concentration in the air was 0.25 µg/m^3 (0.24% TLV). Paraquat was detected in 9 of the 19 spray operators whose urine was analyzed. The maximum paraquat concentration in urine was found in women — 0.76 mg/l, despite the relatively lower exposure. The mean percentage of the toxic dose was calculated based on exposed body-part data. A very low figure was found — 0.05%/h or 0.3%/working d, with a maximum of 0.25% or 1.5% of the toxic dose/d.

Wojeck et al. studied workers exposed to paraquat and diquat during tractor application to tomato and citrus fields and water ways, respectively.[10] Citrus applicators exposed to paraquat averaged 28.5 mg/h with a tank concentration of 0.11%, and 12.2 mg/h with a tank concentration of 0.07%. Urine samples were negative, except in one case which contained 0.033 ppm paraquat in 1 d.

The estimated total body exposure for tomato applicators was 168 mg/h for regular tractors, 26.9 mg/h for enclosed-cab tractors, and 18.38 mg/h for high clearance tractors. The respiratory exposure, calculated by taking an air sample near the breathing zone of the worker, was less than 1% of the total body exposure and ranged from 0 to 0.07 mg/h. Dermal exposure, however, may have been overestimated, because the clothing penetration was not measured.

Chester and Word investigated exposure during the aerial application of paraquat to cotton in California.[8] Dermal and respiratory exposure was considered low. The hourly dermal exposure for a mixer-loader was 0.2 mg/h; for flaggers it was 2.39 mg/h. The highest concentration of paraquat in the air was 26.3 µg/m^3. No respirable paraquat was measured in the breathing zone of any worker. (The highest percentage of respirable droplets was 1.86% of the total amount collected). The conclusion of the authors is that there is no evidence of a toxic hazard to people engaged in the aerial application of paraquat.

Makovskij, cited in EHC 39, studied workers spraying cotton fields with paraquat.[1] Air concentrations varied from 0.13 to 0.55 mg/m^3 air. During arial spraying the paraquat concentrations measured in the air were 4.31 to 10.7 µg/m^3.[9] During harvesting airborn concentrations were 1.245 to 0.516 µg/m^3.

Staif et al. studied paraquat exposure in 35 different situations. A total of 230 analyses of dermal and respiratory exposure pads and 95 hand rinses were performed.[11] Paraquat excretion as an index of absorption was measured in a total of 130 urine samples. The maximum potential dermal and respiratory exposure represented 3.4 mg/h, or 0.06% of a toxic dose; if paraquat was used under the described conditions, almost the whole amount was found in rinses. Respiratory exposure was extremely low — 0.4 to 2.0 µg only in single pads. It was not possible to detect paraquat in urine during and following exposure.

The hazard to children who might place their mouths over yard and garden pressurized despenser nozzles was also evaluated. The highest value for a brief nozzle discharge was 0.53 mg, and for a 1 s discharge — 1.74 mg of paraquat.

Diquat aerosol concentrations range from 0.06 to 0.56 mg/m^3, according to the method of application.[1]

Exposure to paraquat and diquat residue in food is negligible. Environmental contamination does not represent serious problems, because of deactivation and the rapid and complete binding of paraquat to clay products in the soil. Degradation takes place through ultraviolet radiation and soil microorganisms. Chemical degradation is relatively slow.

VII. EFFECTS ON HUMANS

Reports on acute poisoning with paraquat have been summarized.[1] From a total of 925 cases, 356 were fatal. The percentage of death cases vary from 35 to 100%.

TABLE 1
Some Data for Paraquat Intoxications in Pacific Countries

	Fiji	Western Samoa	Papua New Guinea
	1976–84	1981–83	1969–81
Total number of intoxications for the period	237	81	47
Suicide and homocide	not reported	81	16
Accidental and unknown	not reported	not reported	25
Number of death cases	60	not reported	74
Reported cases per year	45	36	15

A. ORAL INTOXICATION

The main route of intoxication is oral. Accidental cases have resulted from incorrectly labeled containers. They usually have a more favorable outcome than suicides. Suicide intoxications are usually fatal. About 30 to 50 death cases occur annually in the U.K. due to paraquat poisoning, mostly suicidal. The situation in Japan is alarming. About 1300 persons die each year from paraquat poisoning.[12]

Fatal paraquat poisoning has become a significant problem during the last 10 to 15 years in some Pacific countries as well. A review of intoxications for the period 1975–1985 has been made by Taylor et al.[13] Both suicidal and accidental ingestions have led to a significant number of deaths. The authors gathered data from published and other reports. The death rate for 100,000 population in Fiji was 2.2 to 8.7. Some data by countries are shown in Table 1.

The target organs for paraquat are the lungs, kidneys, liver, adrenals, skeletal muscles, and heart. With the exception of lung damage (progressive fibroblastic alveolitis), all changes are reversible.[14] Death from paraquat ingestion usually occurs from severe hypoxemic respiratory failure; or later from specific diffuse intraalveolar fibrosis.

The association of recent pulmonary fibrosis, caustic lesions of the digestive tract, and improving tubulonephritis is almost pathognomic for paraquat intoxications in cases of a delayed death.[15] In case of quick death by ingestion of a massive dose, the lesions are acute tubulonephritis, pulmonary damage, and cytolytic hepatitis.[15]

In connection with the severeness and the outcome of the intoxication, three cases of fatal poisoning have been described:[1]

- The most specific for paraquat is pulmonary fibrosis. Death occurs 4 d to several weeks after the absorption.
- A massive dose causes acute fulminant poisoning leading to generalized damage of the lung (edema), kidney (oliguria), liver and adrenal failure. Death usually occurs within 1 to 4 d.
- Less overwhelming poisonings — organ damage and failure — have slower development, and some complications contribute to death: mediastinitis, complication of therapy, etc.

Vale (cited by Crome) described 3 grades of poisoning with paraquat: (1) mild poisoning — less than 20 mg/kg paraquat ingested with gastrointestinal symptoms; (2) moderately severe — 20 to 40 mg/kg taken (11 to 15 ml gramoxone or more than one cachet of granular paraquat) with renal failure and pulmonary fibrosis, most patients died in several days or weeks; (3) severe intoxication — more than 40 mg/kg (15 ml) ingested with multiple organ failure (heart, lungs, liver, kidneys, adrenals, and pancreas), inevitable death followed in a few days.[12]

B. LUNG DAMAGE

Lung damage is characterized by progressive fibroblastic proliferation in the alveolar walls, interstitial tissue, and bronchial epithelium. Later hyaline membrane forms, similar to that found in respiratory distress syndrome.[14] Lung damage becomes apparent long after ingestion. That is why Barnes described paraquat as a "hit and run" poison.[16] The lung damage is nonspecific during the first few days and generally consists of pulmonary edema and hemorrhage. No efficient treatment of the progressive pulmonary damage exists.

Bulivant reported two accidental poisonings by paraquat.[17] They both showed initial symptoms of gastroenteritis. In one case it was followed by hepatic failure and renal insufficiency; in the other it led to respiratory distress and toxic myocarditis. Alveolar fibroblast proliferation and terminal bronchi were found at necropsy.

Davidson and Macpherson described pulmonary damage in three cases of paraquat poisoning.[14] Two of the patients had mistakenly ingested 20% paraquat (one and two mouthfuls and vomited immediately). They developed progressive pulmonary fibroblastic alveolitis on the seventh and sixth day. Chest radiography and respiratory function tests remained normal until day seven and day six, respectively, when granular or confluent opacities became evident. Respiratory deterioration continued; one of the patients died on the 17th day and the other on the 25th. An autopsy found the lungs solid and enlarged. Histologically, almost complete alveoli obliteration by proliferating fibroblastic tissue, emphysema, and inflammatory exudate were observed.

The third case involved the ingestion of a solid preparation of paraquat (45 g 5%). Severe periumbilical pain started 15 min later and the patient was hospitalized and treated effectively with forced diuresis. Urinary paraquat was 14,800 μg/100 ml, and serum was 85 μg/100 ml. After 24 h urine paraquat was 95 μg/100 ml, but the serum level decreased to 0 in 15 min.

Four cases of accidental paraquat poisoning are reported by Vicinovic.[18] One patient developed progressive pulmonary fibrosis and died 19 d after the ingestion. In another, kidney function was impaired, but it recovered in a few weeks.

Very small quantities of paraquat can provoke a lethal intoxication. A 44-year-old man mistakenly ingested gramoxone, which he immediately cast out. In spite of that, a severe intoxication with characteristic symptoms took place, and after 8 d the patient died.[19]

Other cases of poisoning are reported by Larrard, Hargreave et al., Slocombe et al., Jones and Owen-Lloyd, Almog and Siegelbaum, and Bronkhvist et al.[20-25]

One pregnant woman died 15 d after ingesting a teaspoonful of paraquat solution. No changes in the tissues of the fetus were found. The authors explain this by the absence of lung activity and the reduced kidney activity in the fetus.[26]

Almog and Tal reported a fatal case of subcutaneous injection of 1 ml 20% gramoxone solution.[27] After 2 d the person developed transitory right facial paresis. On the third day he complained of chest pain, and his temperature rose to 39°C. X-rays showed a slight infiltration shadow on the right lung base. On the eighth day jaundice appeared, and the liver was 6 cm below the costal margin. The total bilirubin values were raised to 4.6 mg/100 ml (direct 3 mg/100 ml). Serum SGOT rose to 250 U, leucine aminopeptidase was 364 U. These findings disappeared on the 11th day, but the patient developed severe respiratory distress and died on the 18th day.

Toner et al. describe the fine structure of lung tissue, obtained by biopsy, characterizing the early stage of paraquat damage.[28] Edema and fibroblastic proliferation, as well as induration of some alveolar partitions, were found. Biopsy, performed on the seventh day after paraquat ingestion, demonstrated changes in the epithelium of the alveolar septa, determining cell death. In the areas of bare alveolar basal membranes, macrophages and abnormal septal fibroblasts were found. Destruction of the normal lung architectonics was noted, but the alveolar structure was not damaged. The alveoli contained edematous fluid, inflammatory cells, and proliferative fibroblasts with data for collagen deposition, and focal hemorrhages. In some areas, clearly outlined hyaline membranes delineated dilated aerial spaces, and in some places, proliferation of the terminal bronchial epithelium in the surrounding interstitial fibrosis was noted. Histological changes of kidney tubules were established. The lethal outcome of this paraquat intoxication resulted from a progressive lung fibrosis. Parallelly, liver and kidney lesions were also found.

Optical and electron microscopy of the lungs of a young patient, who died 20 d after ingesting a mouthful of paraquat, was performed.[29] Alveoli were extremely rare and contained a variety of cells: red blood cells, macrophages, lymphocytes, and connective tissue cells — all in edematous fluid. These alveoli were lined by granular pneumocytes. Alveolar septi were converted into connective tissue areas with fibroblasts, collagen, elastic tissue, leukocytes, fagocytic cells, and erythrocytes. Capillary permeability seemed to be enhanced by vesicles, forming transendothelial channels or ports, or by the disruption of endothelial cells.

Copland et al. stressed the absence of interstitial fibrosis and vessel growth in one case with paraquat pulmonary damage.[30] The author believes that extensive intraalveolar avascular fibrosis represents a characteristic morphological pattern reflecting a purely alveolar epithelial injury.

Edema and necrotising alveolitis, kidney parenchymal dystrophy, and steatosis in the liver were described in acute paraquat intoxication.[31]

Grozeva reported five lethal cases.[32] Varieties of pulmonal changes have been found in relation to the period of survival.

C. KIDNEY, LIVER, AND GASTROINTESTINAL SYSTEM DAMAGE

Many reported cases of severe paraquat intoxications with clinical picture are dominated by kidney and liver damage, without pulmonary involvement.[17,33,35]

Wright et al. reported 16 patients intoxicated by the intentional ingestion of paraquat.[33] The 6 who took gramoxone liquid preparation died, and the other 10, who ingested weedil (granules), survived. All survivors developed a mild liver and kidney impairment. The high mortality was associated with an excretion rate of over 1 mg/h more than 8 h after ingestion. It is strange that no pulmonary damage has been reported in these patients. Liver and kidney damage were common in all the patients who died.

Increased serum catalase activity was found in patients intoxicated with paraquat due to liver injury.

Mircev reported a paraquat oral intoxication in a 6-year-old boy; he swallowed a mouthful of paraquat by mistake.[35] The illness began with a sudden onset of vomiting and diarrhea. After 5 d of treatment for enterocolitis it became clear that the boy had swallowed a mouthful of gramoxone. The boy was admitted to a hospital 7 d after the injestion in a very grave state with renal failure, anuria, generalized edema, nitrogen retention, intestinal paresis, and metabolic acidosis. The boy was treated successfully with peritoneal dialysis, glucose, levulose, vitamin B_1, vitamin B_{12}, vitamin C, orovit, co-carboxilase, calcium gluconate, aminoacid solution, human protein, plasma, fresh blood, sodium bicarbonate, and ampicillin. This case offers some optimism for the possibility of a favorable outcome, even in advanced, severe intoxication, if adequate treatment is applied; however, Mircev stressed the late development of liver damage (sixth day).[36] Regardless of the preventive treatment and protective means, it was not possible to stop the liver injury.

Vasiri et al. stressed on nephrotoxicity of paraquat.[37] Three patients were poisoned orally and died 17 to 22 d later from respiratory failure; acute renal failure was observed. The glomerular filtration rate improved in 3 weeks. Toxic or ischemic insult, or both, might have been responsible for the kidney damage, according to the authors. Transient proteinuria was observed during the first 2 weeks, marked by aminoaciduria, renal glucosuria, and impaired transport of phosphorus, uric acid, and sodium. These findings are suggestive of proximal tubular lesions.

Renal failure was present on the day following ingestion of 20% paraquat. About 4 μg/100 ml paraquat was found in urine. Treatment with daily hemodialysis improved the renal function by the ninth day.[14]

Fitzgerald et al. reported on kidney damage in paraquat poisoning and treatment.[38] At necropsy, tubular necrosis was found. Fatty degeneration of periportal hepatocytes and sporadic cellular necrosis in the central region of the liver tubules, cholestasis, and portal inflammation have been observed by Shuzuki, 1980, Matsumoto et al., 1980, Takayama et al., 1978.[1]

Central nervous system impairments have also been reported. In severe cases, convulsions, ataxia, and semiconsciousness were observed. Hemorrhagic leukoencephalopathy, hemorrhagic meningitis and demyelination, and focal hemorrhage were found at necropsy.[1]

Only a few incidents have been reported with diquat. The clinical picture was similar, except for lung proliferation or fibroblastic changes.[3] Cerebral damage caused by bleeding in the brain has been reported in one case.

The local effects of 20% paraquat ingestion were reported to appear the next day: the mouth began to feel sore, drinking became more difficult, and on the third day marked ulcerations appeared on the mouth and fauces.[14]

Two cases of esophageal perforation due to paraquat ingestion have been reported by Ackrill et al.[39] The paraquat concentration in plasma was 27 µg/l, in the lungs — 1.5 µg/g, and in the kidneys — 2.6 µg/g.

D. OCCUPATIONAL INTOXICATIONS

Howard performed a clinical survey of workers engaged in paraquat formulation.[40] One group of 18 male subjects from a plant in the U.K. and another group of 18 male workers from a plant in Malaysia were studied. The exposure duration varied from 1.3 to 12.5 years. The medical records were examined. Medical and occupational histories were obtained and skin examinations performed. A number of acute skin rashes were reported (delayed caustic effect), as well as burns, nail damage, and eye injuries, mostly in Malaysian workers due to neglecting the use of protective clothing. Skin damage consisted of erythema and occasionally the formation of bullae. In most cases the recovery was normal. Eye splashes were followed by intense conjunctivitis with blepharospasm and lacrimation. Workers handling solid material experienced occasional epistaxis; they complained about frequent blood spotting of their handkerchiefs after blowing their noses. There was no evidence of any permanent damage to the skin, conjunctivae, and cornea. Vision was not affected. The only chronic effects to be considered are the occupational blepharitis and delayed healing in 3 to 4 workers.

Howard et al. studied the health of 27 paraquat spraymen on a Malaysian plantation using tests for pulmonary function, renal function, and a full hematological screen.[41] No significant differences between the exposed and control group were found.

E. DERMAL ABSORPTION

A small number of fatal accidental intoxications by dermal absorption have been reported. Some of them are by occupational exposure, others are related to the use of paraquat to kill body lice or scabies infestation.[1,42]

There are also cases of pulmonary damage due to percutaneous absorption by healthy persons. Such is the case reported by Jaros[31] and Jaros et al.[43] A farmer prepared a more concentrated solution than recommended, e.g., 8.0 l water and 2.0 l paraquat (10% formulation). He used an old sprayer that leaked freely down his neck, back, and legs. He felt burning on his neck and scrotum after 4 h. In 6 d he developed pulmonary symptoms: a cough, respiratory difficulties. X-rays showed abnormalities on the seventh day. A compensated metabolic acidosis progressed into decompensated metabolic and respiratory acidosis and total respiratory insufficiency. Artificial respiration was initiated. In the blood urea, nitrogen and serum creatinin were raised. He died of renal and pulmonary insufficiency 9 d after exposure, in spite of the applied treatment of corticosteroids, cardiotonics, diuretics, antibiotics, and bicarbonate.

Facial palsy was found to be related to a minute dose of paraquat by dermal absorption.[44]

F. LOCAL SKIN AND NAIL EFFECT

Swam reported two episodes attributable to the local effect of paraquat.[6] One worker developed dermatitis of the scrotum when his trousers were contaminated with diluted spray solution. The dermatitis required 14 d of in-patient hospital treatment. After that, the subject resumed work and further contact with paraquat solution, which produced some itching of the

TABLE 2
Paraquat Concentration in Blood at Different Grade of Intoxication[51]

Intoxication	Paraquat in plasma	Hours
Nonfatal	120 ng/ml	3 h
Nonfatal	50 ng/ml	8 h
Nonfatal	115 ng/ml	4 h
Fatal at 6th d	940 ng/ml	3 h
Fatal at 7th d	420 ng/ml	4 h
14 fatal cases	mean 1263 µg/l	after mean time 8.3 h
26 nonfatal cases	mean 214 µg/l	after mean time 8.6 h

skin. The other episodes were severe and required medical attendance. The person supposedly inhaled some droplets of paraquat while preparing the solution. In the urine, 0.04 ppm paraquat was found. In another trial, 1 case of epistaxis was reported as well.

During 1969, 7 cases of moderate dermatitis, 1 with epistaxis and 7 with conjunctivitis, were reported. Paraquat in the urine was 0.1 to 0.32 mg/dm.[3] More severe dermatitis, such as blistering, ulcerations, and necrosis, were due to excessive contact with concentrated solution.

Positive allergic reaction and positive photopatch response have been reported in a small percentage of exposed persons.[3]

Nail damage followed repeated contamination with diluted paraquat in 55 of 296 sprayers on a sugar estate in Trinidad.[45] The earliest lesions were white or yellow discoloration or transversal bands of white discoloration. In more advanced stages, the sprayers suffered progressive nail deformity with a damaged and irregular nail surface or transverse ridging and furrowing. A grossly irregular deformity of the nail plate was followed by the separation and loss of the nail, associated with subsequent infection. A few months after ceasing exposure, the nails regrew.

Samman and Jonston first reported three cases of nail damage after frequently repeated contamination with concentrated paraquat and diquat.[46]

Ocular damage is possible from splashes of concentrated paraquat that make contact with the eye.[6,40] Stanley et al. reported eye inflammation after the hair was contaminated with reglon.[47] Conjunctivitis was followed by the development of uveitis. Paraquat and diquat produced the burning of conjunctiva after direct splashing.[48,49]

Diquat has effects on nails and mucosae similar to paraquat. Isolated cases of cataracts have been reported in workers exposed to diquat.[3]

G. PARAQUAT CONCENTRATIONS IN BIOLOGICAL MEDIA

Measuring paraquat concentrations could help predict intoxications. According to Proudfoot et al., patients whose plasma concentrations do not exceed 2.0, 0.6, 0.3, 0.16, and 0.1 mg/l at 4, 6, 10, 16, and 24 h, respectively, have good prognosis.[50] This conclusion is based on plasma analysis in 79 patients. The data of Higenbottom et al. support this suggestion (Table 2).[51]

The same authors discussed the dose/pulmonary effect relationship.[51] The authors reported the recent experience of the London Centre of the National Poisons Information Service with 188 cases of paraquat intoxication; 57 died acutely within 7 d and only 12 died later with evidence of pulmonary fibrosis. A patient who had swallowed 2.5% w/v granular form of paraquat (weedol) survived ill effects. The authors concluded that paraquat-induced lung damage was reversible and consisted only in a reduction of gas transfer factor and/or pulmonary edema. With the ingestion of larger doses, lung damage started with edema and developed into fatal pulmonary fibrosis. Acute pulmonary edema was a response to large doses and the patients died in the first 6 d.

A very severe case found high concentrations of paraquat where the death occured 22 h after the ingestion, i.e., 13 mg/1000 g in blood, 68 mg/1000 g in the lungs, 85 mg/1000 g in the liver, 54/1000 g in the kidneys, and 32 ng/1000 g in the heart.[15]

Concentrations in excess of 200 µg paraquat/100 ml urine are an indication of serious poisoning.[52] The highest concentration of paraquat found in the urine of sprayers was 0.32 mg/l, the average being below 0.1 mg/l.[2]

H. TREATMENT OF INTOXICATION[2]

If swallowed, the first step is to induce vomiting. A high fluid intake should be maintained and the patient immediately hospitalized. In case of dermal contamination, careful washing with soap and water is recommended. In the hospital, the most important thing is to remove as much paraquat as possible in order to prevent absorption. Gastric lavage should be performed carefully to avoid damaging esophageal mucosae further. More than 500 ml water solution of 7% bentonite and 10% glycerine final concentration (colloidal aluminium silicate) suspension should be introduced into the stomach within 2 h after ingestion. Fuller's earth 30% can be used instead of bentonite. Forced diuresis, in a patient who had swallowed 45 g 5% paraquat during 24 h, got good results in facilitating paraquat excretion.[53] In case of renal failure, hemodialysis and peritoneal dialysis may be indicated. Oxygen is often necessary, but there is some indication it can contribute to the aggravation of lung damage.

The most important goal is to prevent lung lesions, but up to the present there is no effective method. Immunosuppressive therapy, corticosteroids, prednisolone, cyclophosphamide, beta-bloquants, hypo-oxygenation, and anti-oxidants (vitamin E) have been recommended due to the good results obtained in animals. Some effects of superoxide dismutase were demonstrated in rats only by Autor.[54]

Matthew et al. reported an unsuccessful attempt at transplanting one lung in a 15-year-old boy with severe paraquat respiratory distress.[55] In the transplant, characteristic lung changes were observed, which were attributed to paraquat. The patient died 2 weeks after the transplant.

Prognosis of intoxication is associated with many factors. The most important is the quantity and type of formulation used. Ingestion of more than 6 g liquid formulation is usually fatal, but even in such cases, if treatment is commenced without delay to prevent absorption and renal damage, prognosis could be favorable.

Bismuth et al. reported 28 acute intoxications with paraquat, 11 survived.[4] The authors discussed the prognosis: route of absorption, kind of intoxication (intentional or accidental), dose, stomach status (after meals or empty), and accompanying esophageal damage; in relation with paraquat concentration in plasma. The functional tests of the respiratory system are less in important.

The effectiveness of the treatment is evaluated, as well. Clearing the digestive system through neutralization with different absorbants is of primary importance. Gastric lavage and artificial diarrhea usually come very late. Renal and extrarenal clearance by forced diuresis, hemodialysis, and hemoperfusion are of secondary importance, according to the authors, due to intracellular presence of paraquat in the organism.

Gallow and Petrie reported a patient who survived a severe paraquat poisoning.[52] He ingested a large mouthful of 20% gramoxone. On admission to the hospital, (52 h after ingestion) the initial pulmonary, gastrointestinal, and renal damage had developed. Later, hepatic and myocardial damage occured. The urinary paraquat level was 6800 µg/100 ml. Forced diuresis was initiated with 200 ml of 20% mannitol and a high fluid intake, mainly oral. Only one 5-h hemodialysis was carried out, 67 h after admission. The typical pulmonary changes did not develop.

The authors concluded that forced diuresis helped prevent the development of pulmonary fibrosis. They discussed the prognostic value not only of the quantity of paraquat ingested, but also the rate of removal.

Kean also reported successful treatment of acute intoxication.[56] Williams (cited by Crome) discussed a patient with definite lung damage who recovered with radiotherapy.[12]

I. RECOVERY AFTER TREATMENT OF INTOXICATION

Fitzgerald et al. studied 13 patients (11 adult and 2 children) who survived a recorded episode of paraquat poisoning for at least 1 year.[57] The purpose of the study was to determine the prevalence of residual pulmonary disability. Four adult patients (all smokers) had a mild deficit in pulmonary function. This deficit could be attributed to smoking, but the effect of paraquat was not excluded. One adult patient showed persistent infiltrates and pronounced arterial hypertension. All other patients were considered normal. Based on various radiological studies before and after paraquat poisoning in one patient, the authors concluded that there was some evidence that permanent lung damage could occur accidentally.

Gervais et al. also reported pulmonary sequelae, such as obstructive syndrome, atrophic pneumopathy, emphysema, and interstitial fibrosis with disturbances in pulmonary mechanics and diffusion.[15]

Acute poisoning with diquat (reglon) resulted in less severe clinical symptoms of favorable evolution.[58]

VIII. PREVENTION

A. SAFE LEVELS

The maximum permissible levels for paraquat in the U.S. — 0.5 mg/m^3; for respirable sizes according to ACGIH — 0.1 mg/m^3, (IDLH — 1.5 mg/m^3); in Bulgaria — 0.01[60]; in Hungary — 0.02. For diquat, the maximum permissible level in the U.K. is 0.5 mg/m^3; in the U.S.S.R. and Bulgaria it is 0.1 mg/m^3.[61] The acceptable daily intake for paraquat is 0.001 mg/kg/b.w., and for diquat it is 0.008 mg/kg/b.w.[62]

The ability of paraquat to induce local effects underlines the need for careful handling, as is the case for any caustic material.[40]

Using disposable coveralls and regularly washed impermeable gloves, especially combined with permeable palmar or back ones, would considerably reduce the potential systemic exposure to paraquat and diquat.[10] Persons with skin lesions should not be permitted to work in contact with paraquat and diquat until full recovery.

IX. CONCLUSION

Paraquat and diquat are used as total herbicides. They have high acute toxicity. The mechanism of action is related to single electron reduction, oxidation with flavoproteins and molecular oxygen, and peroxidation of lipid cellular membranes. Paraquat represents a very high hazard due to irreversible lung damage. The most specific effect for paraquat is pulmonary fibrosis. In acute intoxication, hepatic failure and renal insufficiency are also very common. Gastrointestinal and central nervous system impairment may also occur.

Occupational intoxications mostly result from local contamination. Dermatitis (some times severe), epistaxis, conjunctivitis, and ocular damage may be the result of splashes of concentrated paraquat. Nail damage following repeated contamination with diluted paraquat also occurs.

Most important in the treatment of acute poisoning is the prevention of lung lesions, but up to the present there is no sufficiently effective method. Forced diuresis, hemodialysis, and hemoperfusion are important. Immunosuppressive therapy, corticosteroids, etc. are recommended as well.

REFERENCES

1. Paraquat and Diquat, Environmental Health Criteria No. 39, World Health Organization, Geneva, 1984.
2. Paraquat, Data Sheets on Pesticides No 4, Unpublished Document, World Health Organization/Food and Agriculture Organization, 1975.
3. Diquat, Data Sheets on Pesticides, Unpublished Document, World Health Organization/Food and Agriculture Organization, 1979.
4. **Bismuth, C., Dally, S., and Pontal, P.-G.,** Prognostic factors and therapeutic results of paraquat poisoning. Concerning 28 cases, *Arch. Mal. Prof.,* 44, 1, 38, 1983.
5. **Conning, D. M., Fletcher, K., and Swan, A. A. B.,** Paraquat and related bibyridiliums, *Br. Med. Bull,* 25, 245, 1969.
6. **Swan, A. A. B.,** Exposure of spray operators to paraquat, *Br. J. Ind. Med.,* 26, 322, 1969.
7. **Chester, J. and Woolen, B. H.,** Studies on the occupational exposure of Malaysian plantation workers to paraquat, *Br. J. Ind. Med.,* 38, 23, 1981.
8. **Chester, G. and Ward, R. S.,** Occupational exposure during drift hazard during aerial application of paraquat to coffee, *Arch. Environ. Contam. Toxicol.,* 13, 551, 1980.
9. **Seiber, J. N. and Woodrow, J. E.,** Sampling and analysis of airborn residues of paraquat in treated cotton field environments, *Arch. Environ. Contam. Toxicol.,* 10, 133, 1981.
10. **Wojeck, G. A., Price, J. F., Nigg, H. N., and Stamper, J. H.,** Worker exposure to paraquat and diquat, *Arch. Environ. Contam. Toxicol.,* 12, 65, 1983.
11. **Sraiff, D. C., Comer, S. W., Armstrong, J. F., Wolfe, H. R.,** Exposure to the herbicide paraquat, *Bull. Environ. Contam. Toxicol.,* 14, 334, 1975.
12. **Crome, P.,** Paraquat poisoning, *Lancet,* 1, 333, 1986.
13. **Taylor, R., Tama, K., and Coldstein, G.,** Paraquat poisoning in Pacific Islands countries 1975–1985. Tech paper No. 189, South Pacific Commission Noumea, New Caledonia, Dec. 1985.
14. **Davidson, J. K. and Macpherson, P.,** Pulmonary changes in paraquat poisoning, *Clin. Radiol.,* 23, 18, 1972.
15. **Gervais, P., Diamant-Berger, O., Becol-Livesac, J., Guillman, C., and Guyon, F.,** Problèmes médico-legaux et médico-sociaux de l'intoxication aguë par les herbicide du group du paraquat, *Arch. Med. Prof.,* 36, 19, 1975.
16. **Barnes, J. M.,** Poisons that hit and run, *New Scientist,* 38, 619, 1968.
17. **Bullivant, C. M.,** Accidental poisoning by paraquat. Report of 2 cases in man, *Br. Med. J.,* 1, 1272, 1966.
18. **Vucinovic, V.,** Four cases of poisoning with paraquat, *Arch. Hig. Rada,* 29, 261, (in serbo-Croat), 1978.
19. **Oreopoulos, D., Soyannwo, M., Sinniar, P., Fenton, S., McGeown, and Bruce, J.,** Acute renal disease in case of paraquat poisoning, *Br. Med. J.,* 5594, 1, 749, 1968.
20. **De Larrard, J.,** Intoxication agro-chimique par ammonium quaternair (gramoxone), *Arch. Mal. Prof.,* 30, 421, 1969.
21. **Hargreave, T., Gresham, G., Karayanopoulos, S.,** Paraquat poisoning, *Postgr. Med. J.,* 45, 633, 1969.
22. **Slocombe, G., Thorn, P., Toohill, J., and Wood, J.,** A case of paraquat poisoning, *Nur. Times,* 69, 111, 1973.
23. **Jones, G. and Owen-Lloyd, D.,** Recovery from poisoning by 20% paraquat, *Br. J. Clin. Pract.,* 23, 2, 69, 1973.
24. **Almog, C. Y. and Siegelbaum,** Paraquat poisoning in Israel, *Harafuah,* 87, 400, 1974.
25. **Bronkhorst, F., Van Deal, J., and Tan, H.,** Fatal poisoning with paraquat (gramoxone), *Nod. Tüdschr, Geneesk.,* 112, 300, 1968.
26. **Fonelly, J., Gallacher, J., and Carroll, R.,** Paraquat poisoning in pregnant women, *Br. Med. J.,* 3, 5620, 722, 1968.
27. **Almog, C. and Tal, E.,** Death from paraquat after subcutaneous injection, *Br. Med. J.,* 3, 721, 1967.
28. **Toner, P.,** Fine structure of the lung lesion in case of paraquat poisoning, *J. Pathology,* 102, 3, 182, 1970.
29. **Dearden, L. C., Fairshter, R. D., McRae, D. M., Smith, W. R., Glaser, F. L., and Wilson, A. F.,** Pulmonary ultrastructure of the late aspects of human paraquat poisoning, *Am. J. Pathol.,* 93, 667, 1978.
30. **Copland, J., Colin, A., and Shulman, H.,** Fatal pulmonary intraalveolar fibrosis after paraquat ingestion, *N. Engl. J. Med.,* 291, 8, 290, 1974.
31. **Jaros, F.,** Acute percutaneous paraquat poisoning, *Lancet,* 1, 275, 1978.
32. **Grozeva, M.,** Pathomorphology of the Lung Damage Due to Isolate or Combined Intoxication with Paraquat and Ethanol and Its Importance for the Forensic Medicine, Ph.D. thesis, Sofia, Bulgaria, 1985.
33. **Wrigth N.A., Yeoman W.B. and Hale K.A.** Assessment of severity of paraquat poisoning. *Br. Med. J.,* 2, 396, 1978.
34. **Fisher, H. K., Humphries, M., and Bails, R.,** Recovery from renal and pulmonary damage, *Ann. Intern. Med.,* 75, 731, 1971.
35. **Mircev, N.,** A case with gramoxone intoxication in childhood, *Hig. Zdraveop.,* 30, 3, 40, 1987.
36. **Mircev, N.,** Acute intoxications with gramoxone (paraquat), *Vutr. Bol.* (Internal Diseases), 16, 6, 99, 1977.
37. **Vaziri, N. D., Ness, R. L., Fairshter, R. D., Smith, W. R., and Rosen, S. M.,** Nephrotoxicity of paraquat in man, *Arch. Inter. Med.,* 139, 172, 1979.

38. **Fitzgerald, G. R., Barnivill, G., Silke, B., Carmody, M., and O'Dwyer, W. F.,** The kidney in paraquat poisoning, in Proc. Europ. Dialysis and Transpl. Association, Helsinki 1974.
39. **Ackrill, P., Hasletone, P. S., and Ralston, A. J.,** Oesophageal perforation due to paraquat, *Br. Med. J.,* 1, 1252, 1978.
40. **Howard, J. K.,** A clinical survey of paraquat formulation workers, *Br. J. Ind. Med.,* 36, 220, 1979.
41. **Howard, J. K., Sahopathy, K. N., and Whitehead, P. A.,** A study of the health of Malaysian plantation workers with particular reference to paraquat spraymen, *Br. J. Ind. Med.,* 38, 110, 1981.
42. **Binns, C. W.,** A deadly cure for rice, *Papua New Guinea Med. J.,* 19, 105, 1976.
43. **Jaros, F., Zuffa, L., Krationova, R., Skala, I., and Domsova, A.,** (Acute percutaneous gramoxone intoxication), *Pr. Lék.,* 7, 260, 1978, (in Czechoslovakian).
44. **Mourin, K. A.,** Paraquat poisoning, *Br. Med. J.,* 4, 486, 1967.
45. **Hearn, C. E. D. and Keir, W.,** Nail damage in spray operators exposed to paraquat, *Br. J. Ind. Med.,* 28, 399, 1971.
46. **Samman, P. D. and Jonston, E. E.,** Nail damage associated with handling of paraquat and diquat, *Br. Med. J.,* 1, 818, 1969.
47. **Stanley, C. and Lewis, D.,** Ocular damage by paraquat and diquat, *Br. Med. J.,* 224, 2, 5599, 1968.
48. **Smith, M.,** Paraquat summary of experience of human exposure, ICI Special Centre Toxicology Bureau, January, 1966.
49. **Cant, J. and Lewis, D.,** Ocular damage due to paraquat and diquat, *Br. Med. J.,* 3, 59, 1968.
50. **Proudfoot, A. T., Stewart, M. S., Levitt, T., and Widdop, B.,** Paraquat poisoning: significance of plasma-paraquat concentrations, *Lancet,* 2, 8138, 330, 1979.
51. **Higginbottom, T., Crome, P., Parkinson, C., and Nunn, J.,** Further clinical observations on the pulmonary effect of paraquat ingestion, *Thorax,* 34, 161, 1979.
52. **Galloway, D. B. and Petrie, J. C.,** Recovery from severe paraquat poisoning, *Postgrad. Med. J.,* 48, 684, 1972.
53. **Korr, F., Patel, A., Scott, P., and Tompsett, L.,** Paraquat poisoning treated by forced diuresis, *Br. Med. J.,* 3, 290, 5613, 1968.
54. **Autor, A. P.,** Reduction of paraquat toxicity by superoxide permutase, *Life Sci.,* 14, 1309, 1974.
55. **Matthew, H., Logan, A., Woodruff, M. F. A., and Heard, B.,** Paraquat poisoning. Lung transplantation, *Br. Med. J.,* 3, 759, 1968.
56. **Kean, M. W.,** Recovery from paraquat poisoning, *Br. Med. J.,* 2, 292, 5613, 1968.
57. **Fitzgerald, G. R., Barniville, G. Gibney, R. T. N., and Fitzgerald, M. X.,** Clinical, radiological and pulmonary function assessment in 13 long-term survivors of paraquat poisoning, *Thorax,* 34, 414, 1979.
58. **Oreopoulos, D. and Ecoy, J. M.,** Diquat poisoning, *Br. Med. J.,* 1, 749, 5594, 1961.
59. NIOSH pocket Guide to Chemical Hazards, U.S. Department of Health and Human Services, September 1984.
60. **Bainova, A. and Vulcheva, V.,** Experimental substantiation of gramoxone MAC in working areas, in *Works of the Research Inst. of Hygiene and Labour Protection, 23,* Med. i Fiskultura, Sofia, 1972.
61. **Bainova, A. and Vulcheva, V.,** Experimental substantiation of Reglon MAC in the working environment, in *Problemi na Higienata, 3,* Med. i Fiskultura, Sofia, 1977, 11, (in Bulgarian).
62. Guide to codex recommendations concerning pesticide reesidues. Part II. Maximum limits for pesticide residues, Food and Agriculture Organization/World Health Organization CAC/PR-2, Rome 1986.

Chapter 12

MISCELLANEOUS PESTICIDES

There is very limited information about the human effect of many groups of pesticides in use. This chapter discusses some pesticides from different groups (Table 1).

From these groups, dinitroorthocrezoles (DNOC) cause severe intoxications. They have an expressed hepatotoxic effect and act as blocking agents on the oxidative phosphorylation.

Sovljanski et al. reported four lethal cases of acute DNOC intoxications.[1] Two resulted from a failure to use protective means while spraying fruit trees. In all four cases, death was caused by asphyxia. The authors supposed that DNOC had a local effect on the swallow reflex.

A mixture of DNOC and dinitrobutylphenol caused the death of an agricultural worker. Death occurred 9 d after exposure and 4 d after the symptoms appeared.[2] However, this case study was not adequately investigated.

A 30-year-old worker without liver damage sprayed fruit trees one afternoon using a water solution of dinitroorthocresole. Working without mask, his clothes, hands, and hair were contaminated with dust. The next morning his temperature rose up to 39°C; he thought he had a cold. On the fourth day he developed arrhythmia and pseudojaundice. The examination showed data for moderate jaundice, cytolysis, and hypercholesterolemia, as well as elevation of total bilirubin and transaminase activity. Two months after the accident the patient recovered.[3]

The herbicides of the halogenated hydroxybenzonitrile group (ioxynil, bromoxynil) and dinitroartocresol share a similar mechanism of action (Table 1). They also are decoupling agents to the oxidative phosphorylation.

Conso et al. reported acute occupational intoxication with oxanyl after a combined herbicide (also containing isoproturan and mecaprop) in water solution was used.[4] The clinical picture was dominated by hyperthermic sweating (40°C), pains in the joints, thirst, gastrointestinal disturbances, and headaches. Hepatomegalia and icterus developed after 48 h. Blood and urine analysis revealed anemia, increased total bilirubin (22 mg/l), mostly free bilirubin, and 5 to 8 times increased transaminase; alkaline phosphatase and coagulation were normal. They found lung edema and cardiomegalia on the fifth day. Treatment with antipiretics and hydration was successful, and 2 months later the patient recovered completely.

Herbicides from the group of triazine usually have a local irritative and allergic effect.

Chateau et al. reported contact dermatitis due to simazine use in agricultural workers.[5] A pronounced eczema had been persisting for two consecutive years and needed hospital treatment. After recovery, the farmer did not have contact with simazine, but the eczema reappeared due to the large use of this herbicide by his neighbors. Only air transport of simazine was believed to have produced the reappearance of the allergic reaction. This observation demonstrates the high allergic potential of triazines.

Intoxication with monofluorescent acid is manifested by CNS stimulation, convulsions, midriasis, bradycardia, hypotonia, hypoglycemia, heart disturbances, and coma. Treatment with acetamide is considered specific.[6]

Iwasaki et al.[7] described cardial disturbances from intoxication with organofluorine compounds (nizole).

TABLE 1
Properties of Some Herbicides

Structural formula	Names and properties

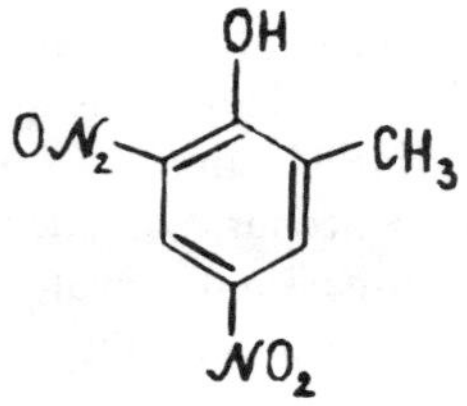

Structural formula (column header)

Names and properties (column header)

Dinitroortocresol (DNOC)
Solid, yellow, m.p. 86°C,
solubility in water at 15°C – 0.013%
LD_{50} 20–50 mg/kg

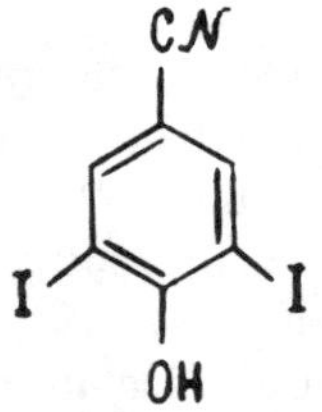

Toxinyl, tortil
Colorless solid, m.p. 212°C,
LD_{50} 110 mg/kg

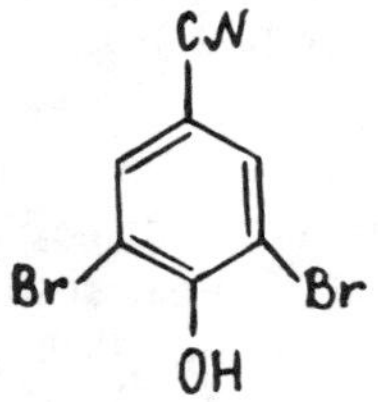

Bromoxyl
White odorless solid, m.p. 194°C,
insoluble, LD_{50} 260 mg/kg

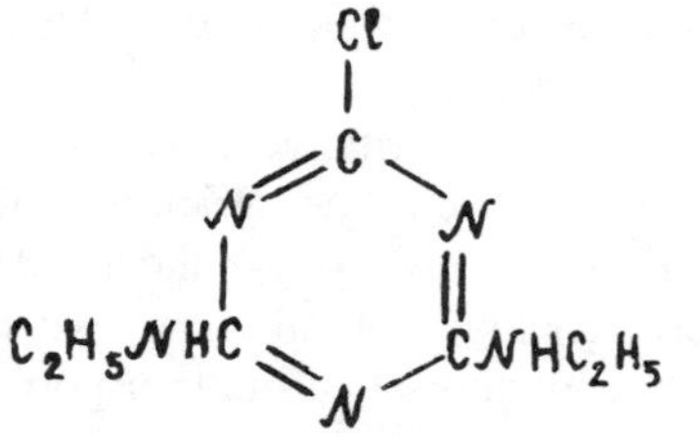

Simazine
Solid, m.p. 225,
solubility in water 3.5 ppm at 20°C,
LD_{50} > 5 g/kg

REFERENCES

1. **Solvjanski, M., Popović, D., Tasic, M., and Solvjanski, R.,** Intoxications with dinitroorthocrezol, *Arch. Hig. Rada Toxicol.,* 22, 329, 1971.
2. **Heyndrickx, A., Mayes, R., and Tybergheyn, F.,** Fatal intoxication by man due to dinitroorthocrezol (DNOC) and dinitrobutylphenol (DNBP), Overduck unit de Mededelingen van de Landbrokhogeschool an de Opzoeking-stations van de stasi de gent. DEEL XXIX, 3, 1964.
3. **Prost, G., Vial, R., and Tolot, F.,** Dinitroorthocresol poisoning involving the liver, *Arch. Mal. Prof.,* 34, 556, 1973.
4. Conso, F., Gibaud, G., Girard-Wallon, C., Alix, M., and Proteau, S. J., Intoxication professionelle par une preparation herbicide contenant de ioxynil, Arch. Hyg. Travail, 585, 1973.
5. **Chateau, M. S., Parant, C., Larche-Mochel, M., and Doignon, J.,** Les dermatoses professionelles dues aux triazines. A propos d'un cas d'eczème de contact à simazine, *Arch. Mal. Prof.,* 43, 8, 659, 1982.
6. **Hashimoto, Y., Makita, T., Mori, T., Nishibe, I., Nogushi, T., and Watanobe, S.,** Preclinical experiments on the antidote against monofluoracetic acid derivatives poisoning in mammals, *Whither Rural Med.,* Proc. 4th Int. Congr. Rural Med., Usuda (Tokyo), 1969, 1970, 59.
7. **Iwasaki, I., Nawa, H., Hara, A., Takagi, S., and Hyodo, K.,** Agricultural organofluoride poisoning, *Fluoride,* 3, 3, 121, 1970.

Chapter 13

COMBINED EXPOSURE

I. INTRODUCTION

In agriculture and public health, different pesticides are very often applied simultaneously during the same season. Therefore, a combined effect could be expected from the combined action of the pesticides.

The clinical picture of such an intoxication has not yet been clarified, and some authors even reject the possibility of diagnosing it. The difficulty comes from the fact that at a long-term exposure to small quantities of pesticides, injuries may be difficult to differentiate from impairments of other etiology.

The general population is also subjected to a combined pesticide exposure from environment and food. Murphy et al. analyzed the results of a 4-year study on blood and urine pesticide residues or metabolites in the general population.[1] About 28,000 persons, aged 6 to 74 years, from 64 communities in the U.S. were studied.

Total DDT was found in 99% of the population, pentachlorphenol in 79%, dimethylphosphate in 12%, and beta-benzenehexachloride in 14%. About 25 other residues were found in lesser amounts.

The toxicological significance of pesticide residues in the organism is not clear. Sometimes they are accompanied by isolated biochemical changes whose toxicological implications should be carefully evaluated; they could be indicating either an adverse effect or only an enzyme short- or long-term adaptation.[2,3]

False positive associations of exposure and outcomes are inevitable; a well matched control group is not easy to find. In this respect the work of Warnick and Carter represents a definite interest.[4] They performed a 4 year study on workers occupationally exposed to various pesticides. They used about 50 tests to check the different organs and body systems in 70 men exposed to pesticides and 30 controls. Slight differences were found, but they were not in the same direction in the different studies. Exposure seems to increase variance in the values since nine of the tests had significantly greater standard deviations for the exposed group. The study is not conclusive concerning the existence of a chronic effect from a moderate exposure to pesticides.

Kaloyanova performed a 2 year epidemiological study on 200 agricultural workers and 100 controls.[5] Some differences were found between the 2 groups. Significantly, more frequent disturbances of the nervous system (neuroses), respiratory system (bronchitis and emphysema), and the gastrointestinal tract and liver, as well as of the cardiovascular system (hypoxia of myocardium and hypertension) were observed in the exposed group. Among the laboratory investigations the more frequent leukopenia, raised bilirubin, reduced serum albumin and increased alpha-globulins should be noted, as well as the activation of some phagocytosis functions. These changes do not have an essential clinical value for the single individuals, but they are characterizing the process.

Under a combined pesticides action, manifestations of antagonism are posssible, which should decrease the adverse effect. It is well known that prior exposure to organochlorine compounds prevents, to some extent, the effect of organophosphorous compounds: in rats, pretreatment with aldrin increases the lethal dose of parathion sevenfold. A ChEA increase, provoked by organochlorine compounds, is a possible mechanism of this antagonism with organophosphorous compounds.[6] At the same time, manifestations of potentiation may also develop, at work with several organophosphorous pesticides.

Dinman reported dichlorphos and chlordane intoxication in a chemist who had worked without observing the necessary preventive measures.[7] Dinman noted a greater inhibition of CNS symptoms, as well as a more rapid recovery of erythrocyte ChEA compared to that of plasma.

Steigalo et al. have observed the intoxications of children in Kirgisia.[8] The children consumed fruits treated with a mixture of DDT and chlorophos. Three of these intoxications developed comparatively mildly; changes in the internal organs were not established by the objective examination. The rest of the children had pronounced intoxications: gastroenteritis was observed and, on the eigth day, symptoms of myocarditis and hepatitis. Blood changes were not noted, and after 2 to 3 weeks the manifestations of intoxication faded away. A repeated examination 5 months later detected dull cardiac sounds and spleen painfulness in only a few children. Blood ChEA was significantly increased up to the end of the first month after intoxication. The authors attributed the increased ChEA to the combined effect of the two compounds. According to them, DDT compensated for the inactivating effect of organo-phosphorous compounds and led to ChEA increase.

Severe intoxications due to combined exposure to organophosphorous and organochlorine substances have also been reported. Ingestion of 200 cm^3 diazinone and DDT mixture has caused tachycardia, cramps, dehydration, and hyperthermia up to death on the 10th day. The possibility of a long-term action of the diazinone metabolites and the existence of a synergism between the products of degradation of DDT and diazinone has been assumed.[9]

The existence of a toxic potentiation between EPN and malathion has been proved. EPN inhibits the enzyme system of malathion detoxication. These facts are valid for many other organophosphorous formulations.

Single cases of intoxication with rich symptomatics at combined exposure have been reported. Soubrier et al. described 4 cases of poisoning due to the application of disinfectants and insecticides.[10] The first case was a worker who had been engaged in disinfection and desinsection for 20 years; he developed the following symptoms: headaches, dizziness, insomnia, weakness, and increased excitability. The second case involved a 2-year exposure resulting in headaches in the frontal area, lumbal pains, lack of appetite, asthenia, nerve excitability, and memory disturbances mainly after DDT spraying; all the symptoms disappeared after changing the work. The third case, after a 1-year exposure, had headaches, itching in the eye area, poor sleep, and dizziness. The fourth case, after 1 to 2 years of exposure displayed dyspnea attacks of an asthmatic type, usually after spraying organophosphorous compounds or working with formaldehyde.

Exposure to parathion, DDT, and phosdrin twice a week in the course of 2 years has provoked muscle weakness, abdominal pains, myosis, muscular dystonia, and increased salivation (Brachfeld and Zavon).[11]

Parathion and cupric sulphate, sprayed in the course of 1 d by the same agricultural worker, has provoked gastroenterological, respiratory, and neurological disturbances with partially affected consciousness (Thiodet et al.).[12]

Often exposure to pesticides is influenced by physical factors. Pesticides applied with tractor sprayers exposed to heat stress result in more frequent health deviations, especially in the nervous and cardiovascular systems, the intestines, liver, etc. These changes have been attributed to the complex action of the chemical and physical factors.[13]

Based on the available literature about the health effects of a great number of pesticides, we attempted to group these effects by organs and systems.

II. RESPIRATORY SYSTEM

The effects of pesticides on the respiratory system are reported by Lings, Barthel, Lehnigk et al., etc.[14-16]

Lings studied a group of fruit growers who used a total of 92 different pesticides, with the active ingredients of 156 spray formulations.[14] Each individual was in contact with 13 substances on the average (3 to 27), the most often used being captan, paraquat, parathion, azinphosmethyl, diquat, amitrol, benomyl, and simazin.

The author interviewed 181 workers and performed the Wright peak flow meter test and an X-ray examination of the chest. Symptoms related with spraying were present in 40%, lung symptoms (coughing, expectoration, dyspnea, or a combination) were in 19%. The Wright test was administered to 178 persons — in 18.5% the peak flow was reduced by more than 10%. Relevant X-ray findings (accentuation of pulmonary markings, infiltration) were observed in 23.8%. The author compared the symptoms and the positive findings of the paraquat users with those of nonusers. People exposed to paraquat suffered more frequently and had lung effects, although the statistical significance was low. As an indirect control group, 132 farmers were interviewed. The questionnaire responses showed that the frequency of symptoms was not significantly higher among those who had used pesticides (biocides).

A higher percentage of farmers demonstrated breathlessness, while fruit growers had more frequent coughing and expectoration, as well as headaches. The author did not perform the Wright peak flow meter test and X-ray examination on any farmer, which made the interpretation of the results difficult. Nevertheless, he concluded that exposure to biocides seemed to cause lung disease, which was characterized by the following principal features: pneumonia (more or less transient round infiltration), and chronic progressive pulmonary fibrosis.

Lehnigk et al. used an adequate control group in their study.[16] They found respiratory function disturbances in workers exposed to pesticides.

Barthel underlined the need for epidemiological investigations of pesticidal aerosol effects on respiratory systems.[15]

III. LIVER AND KIDNEYS

The International Workshop on Epidemiological Toxicology of Pesticide Exposure concluded that the most promising tests for assessing the liver function in people exposed to pesticides are: alkaline phosphatase, aldolase, ornithinecarbamoyltransferase, SGOT/SGPT ratio, and the LDH ratio.[17]

Stoyanov found about 53 to 59% increased SGOT and SGPT in the pilots and aeromechanics he studied.[18] AF changes were present in 6 to 9%. In 44% of the investigated people, an increased gamma-glutamyl-transpeptidase activity was observed. However, the lack of an adequate control group makes the interpretation of the results difficult.

Petkova and Jordanova[19] found increased aspartate-aminotransferase and ornitinaminotransferase, decreased cholinesterase, as well as depressed alkaline phosphatase and glutamyltranspeptidase in people exposed to different groups of pesticides, mainly OP, OCP, dithicarbamates, carbamates, etc. The authors consider these findings a demonstration of nonspecific, subclinical, reversible functional changes in the liver.

Bezuglyi and Kaskevich reported disturbances in the liver functions of agricultural pilots.[20] They found more frequent changes in serum proteins. Hypercholesterinemia and hyperbilirubinemia were found in young persons. All tests in the control group were negative. The authors considered these findings an indication of the hepatotoxic effect of pesticides.

Trendafilova et al. reported changes in the lipid metabolism of greenhouse workers; betalipoproteins and total phospholipids were increased.[21] The changes were more pronounced in workers with a shorter duration of service. The authors suggested that an adaptation takes place with increased duration. Zolotnikova recommended using histidase activity in the blood serum as an early indicator of hepatopathies due to pesticide exposure.[22] She found increased histidase in 69% of the 40 workers investigated; cholinesterase activity in serum was inhibited in 58% of the examined workers. Increases in alanilaminotransferase, histidase, alkaline phosphatase, and bilirubine were found in aeromechanics. More frequent impairments of the respiratory organs, skin, and subcutaneous tissue, as well as radiculitis, were observed.[23]

Bezuglyi and Komarova reported functional hyperbilirubinemia (syndrome of Giebery) in workers exposed to different groups of pesticides in greenhouses, orchards, and vineyards.[24] They studied 228 persons with long-term exposure. In 16 persons (7%), the syndrome was present. Bilirubin levels of 20.5 to 34.2 μmol/l were measured. The transaminases and proteins in the serum were normal.

Giebery syndrome (increased indirect bilirubin without hemolysis and slightly expressed icterus) was accompanied by asthenia, higher fatigability, sweating, dryness in the mouth, and pains in the right subcostal area; the liver increased about 1.5 to 2.5 cm. Phenobarbital, as an inductor of microsomal enzymes in hepatocytes, contributed to the reduction of bilirubin down to the norm. In subjects having long-term contact with pesticides, mainly organophosphorous and organochlorine, disturbances in some metabolic processes were found. The authors underlined the need to observe an appropriate diet to normalize the oxidation reduction processes.

Tocci et al. studied several parameters of liver and kidney function, as well as amino acid levels in the blood of people exposed to large amounts of pesticides.[3] The results indicated changes in 30% of the people studied, namely high levels of several amino acids in the plasma, higher SGOT, alkaline phosphatase, increased serum osmolarity, and creatinine. The phosphorus reabsorption index decreased in 16% of the workers. The authors recognized the difficulty in assessing these effects which might have been related to either cellular damage or beneficial cellular adaptation.

The negative findings of medical observation of people working with pesticides are also of interest, and their publishing has been encouraged. A special investigation on kidney function in persons with long-term exposure to pesticides was performed by Morgan and Roan.[26] They observed 65 subjects, agricultural workers and employees of a factory that produces pesticides, as well as 25 control subjects; 14 had manifested symptoms of intoxication in some periods prior to the investigation. They were exposed to DDT, organophosphorous compounds, and toxaphen. During the last 6 months they all were healthy. The investigation showed that their kidney function was not disturbed.

Gombert et al. did not find differences in the glomerular filtration or the index of phosphate resorption in exposed subjects or control subjects.[27]

IV. IMMUNOTOXICITY

Pesticides may exert a deleterious effect on lymphoid organs, and in this way disturb the immune response of delayed and nondelayed types. They may reduce the nonspecific resistance of the organism to infections, spontaneous mutations, and malignant cells. Some pesticides are haptens, capable of binding to proteins to form complex antigens. This leads to allergy development, which can be demonstrated as contact dermatitis, rhinitis, bronchitis, bronchial asthma, pharyngitis, and allergic diseases of other organs and systems.[28]

A. ALLERGY TO PESTICIDES

Allergic diseases in agricultural workers are often related to pesticides.[29,30]

Vasher and Vallet published a study on the frequency and etiology of allergies supposedly caused by pesticides and fertilizers used in France for 10 years.[31] Out of a total of 9864 allergies, 690 (14%) could be attributed to pesticides and fertilizers. According to the clinical picture, the authors arranged them in the following order:

- Allergies with predominating skin manifestations, 511 cases including 412 cases of contact dermatitis and 89 cases of atopic and other dermatitis (urticaria and angioneurotic edema of Quincke, simple erythema nodosum, pigment erythema, nonthrombopenic purpura, and erythrodermia)
- Respiratory allergy, 117 cases (including asthma in 17% of the cases)
- Blood manifestations, 15 cases, e.g. isolated eosinophilia, hemolytic anemia, leukopenia, agranulocytosis, and thrombopenic purpura
- Serum disease, 8 cases
- Anaphylactic shock (as an exception), 4 cases
- Otorhinolaryngologic disturbances, 28 cases (rhinitis, sinuitis, laryngitis, otitis)

Cardiovascular, neurological, gastrointestinal and endocrine disturbances attributed to allergies without any toxic manifestations have been found after long-term exposure to pesticides, and they are cautiously discussed by the authors.

By means of the "removal-repeated accepting" test the authors established: manifested occupational allergies in 11 cases; symptoms of undoubted occupational origin, but without any possibility to identify the allergen, in 15% of the cases; polysensibilization of predominantly occupational character (low significance of nonoccupational allergens) in 40%; polysensibilization with predominant nonoccupational character in 16%; and exclusively nonoccupational sensibilization (complications due to to occupational stimulation) in 16% of the investigated. The authors presented a detailed table of the allergies provoked by different pesticides and fertilizers.

Dermatitis due to sensibilization from multiple contacts with organochlorine, organosulphurous, organophosphorous, organomercurial, and other pesticides was described by Kambo et al.[32]

Kozintseva et al.[33] studied the possible formation of circulating antipesticide antibodies as evidence of heteroimmune response to the effect of tetramethyl thiuramdisulphide, DDT, chlorophos, and ziram. The data showed that serum antibodies were generated in 20 of the 24 subjects. The authors found that the intensive generation of antibodies induced by these pesticides correlated with the results of other tests evidencing the immunological phase of the allergic process. Using serological reactions as an additional test to confirm the allergic character of the diseases with chemical etiology was proposed.

Popov et al. found latent allergy in 31.68% of the women working in greenhouses.[34] The most common allergen was acrex (11.8%). A polyvalent allergy to different pesticides was also established.

Karimov studied skin sensitivity of 233 cotton growers to different pesticides.[35] In 27 (4.2%) of 633 samples, he found positive reactions to OP pesticides.

B. IMMUNOSUPPRESSION

Katsenovich et al. studied the immune status of patients with different degrees of combined insecticide intoxication.[36] Modern methods were used to assess the T- and B-system of immunity. The quantitative deficit in T-lymphocytes was accompanied by a reduction of their functional activity. The number of B-lymphocytes was increased in most patients. The ratio of T- to B-lymphocytes was altered to varying degrees, depending on the grade of intoxication. Some suggestions were made for developing an autoimmune processes.

Polchenko and Zinchenko studied 12-year-old children living in rural areas of limited or intensive pesticide application. A significant decrease of the titers of isochemaglutinins and immunoglobulins G, A in serum, and particularly of immunoglobulin A in saliva, was found.

Differences in T-lymphocytes were also found. The authors explained the increase in most of the respiratory diseases of the exposed children as disturbances in the immune defense mechanisms.

Zolotnikova reported disturbances in the immunological reactivity of persons working with pesticides in greenhouses.[38] An increased quantity of skin microflora was found, as well as a decrease in the phagocytic activity of leukocytes. Changes in the humoral factors of immunity were less expressed, e.g., the titration index of antibodies and hemolytic activity of serum.

Allergic skin diseases are more frequent in people who worked from 1 to 5 years, while bronchial asthma, liver and kidney damage, and vegetative impairments are more frequent in workers with over 5 years service.

Bezuglyi reported, from clinical and epidemiological studies, more frequent nonspecific diseases of the cardiovascular, nervous, and hepatobiliary systems, gastrointestinal tract, female sexual organs, and the eyes (conjunctivitis and blepharitis) in persons subjected to long-term exposure to pesticides.[39]

Surveys on the general population of agricultural regions or on people using household pesticides show higher general morbidity. They also show an increase in diseases of the respiratory ways (asthma, sinuitis, and bronchitis), leukemia, conjunctivitis, and blepharitis (Chezzo et al.).[40,41]

C. HEMATOLOGICAL CHANGES

A 4-year longitudinal epidemiological study was performed by Romash and Dorofeev[42] on agricultural workers exposed to OCP, OP, phenol nitroderivatives, copper organic compounds, etc., in order to determine the effect of pesticides on peripheral blood. Workers from a food production plant were investigated as a control group. The most demonstrative indicators were considered to be a decreased hemoglobin content, an erythrocyte count showing morphological deviations, neutropenia, lymphocytosis, eosinophilia, granulation, and vacuolization in the cytoplasm. Cytochemical reactions showed increased fetal hemoglobin in 21% of the cases and increased the quantity of ciderocytes.

Decreased myeloperoxidase activity, increased alkaline phosphatase activity in neutrophiles and succidehydrogenase and lactatodehydrogenase in lymphocytes were found.

Similar results were reported by Sutneeva.[43] She found decreased hemoglobin, erythropenia, leukopenia with neutropenia, and eosinophylia.

Tsvetkova et al.[44] studied cation proteins (low molecular basic proteins related to the enzyme function) in the granulocyte lysosomes of agricultural workers exposed to pesticides. A significant reduction of cation proteins was established compared to the control group, which was better manifested during the active pesticide application period. These findings are believed to indicate disturbances in cellular metabolism, decreased synthesis processes.

Long et al. studied the amount of pesticides used and the blood and urine values.[45] The total pesticide applied by the high use group and the erythrocyte sedimentation rate showed a negative correlation in the respective subjects. Positive correlation was established between the total pesticide applied and the blood uric acid, as well as bilirubin 1-min values. The authors discuss these results with caution and suggest further studies.

D. NERVOUS SYSTEM

Pesticides may provoke functional or morphological disturbances in the central, peripheral, and vegetative nervous systems.

Usually, nervous system effects are connected with specific compounds (organomercurials, organophosphates, etc.) which produce specific effects. Only single publications related to combined exposure.

TABLE 1
Visual Disturbances in Workers Exposed to Pesticides

Type of disturbances	Frequency of disturbances %	
	Exposed workers	Control group
Decrease of corneal sensitivity	23.0	5.8
Impaired color discrimination (chrom)	20.3	8.7
Narrowing of visual field (achrom)	58.2	14.4
	64.4	
Enlargement of the blind spot		14.2
Reduced adaptation	16.6	14.2

Fokina studied winegrowers and gardeners exposed to copper-containing OP, carbamates, OCP, etc.[46,47] She found an increased frequency of nervous system diseases, manifested in vegetative vascular dystonia, asthenic vegetative syndrome, hypothalamic syndrome, and diseases of the peripheral nervous system. These persons also showed more frequent cerebral vascular diseases (cerebral atherosclerosis and hypertension). Rheoencephalographic estimations demonstrated decreased blood input in different brain areas of exposed persons. An adequate control group was used to compare the findings. The statistically significant changes were only discussed.

Roder and Lindner studied agricultural workers exposed to OCP, OP, dithiocarbamates, aromatic nitrocompounds, etc.[48] In 15 of 51 persons, subjective symptoms indicating neuropathy were present. Decreased nerve conduction velocity was demonstrated on the peroneal nerve.

E. VISION

In the last few years increasing attention has been paid to eye impairments and visual abnormalities caused by pesticide exposure. Since the eye is a highly sensitive organ that reacts rapidly to irritating stimuli, especially occupational ones, Kalič-Filipovič supports some ophthalmological indices for early diagnosis of occupational intoxications.[49] An improved preventive control is proposed, including a more systematic investigation of the combined effect of pesticides, in which ophthalmological investigation should take its place. The author found that 38.8% of the subjects employed in the pesticide production he investigated lost color-discriminating ability (compared to 8% in the control group); in 25% of the cases, the subjects suffered lower dark adaptation.

Nuritdinova performed ophthalmological studies on 638 workers, aged 20 to 50 years, who had from 1 to 8 years occupational exposure to phosphorous and organochlorine compounds.[50] In 80% of the subjects, symptoms of chronic intoxication have been established (sympathicotonia, asthenia, polyneuritis, and encephalopathy). Eye changes were found in 11% of the cases. Allergic conjunctivitis was found in 33 cases, blepharitis in 6, retinal angiopathy in 28, and retrobulbar neuritis in 2.

Glazko investigated 260 agricultural workers chronically exposed to pesticides and complaining of eye disturbances (Table 1).[51] Clinical investigation revealed symptoms of chronic intoxication. Neuroophthalmological manifestations were more frequent among those who had had symptoms of chronic intoxication by pesticides with systematic effect.

Fournier described neurobulbar neuritis, provoked by inhalatory exposure to organochlorine and organophosphorus compounds.[2] Campbell described neurobulbar neuritis after exposure to pentachlorphenol, dichlorobenzol, and DDT (used to control insects in wooden furniture).[52]

Imazumi observed conjunctivitis, hypervascularization, and corneal ulcerations as a local effect of pesticides.[53]

According to Sugaya, eye injuries due to pesticides in Japan are as frequent as those due to trachoma.[54]

In 1966 California, 25% of the occupational diseases due to pesticide exposure involved the eyes.

Trusiewiczowa reviewed the effects of pesticides on the visual system.[55] She summarized the most frequent changes following long-term exposure: myopia, retinal degeneration, optic nerve damage, accomodation spasm, and glaucoma. She underlined the necessity of more studies in this field.

V. LONG-TERM EFFECTS

The data on teratogenicity, mutagenicity, and carcinogenicity of pesticides in animals are numerous; human data are rather scarce. Data on individual chemicals are given under the respective chapters. Here, only a few publications dealing with long-term effects of combined exposure should be mentioned.

Joder et al. studied lymphocyte cultures from 42 pesticide applicators and 16 control.[56] They found a marked increase in the frequency of chromatid lesions. This phenomenon was especially pronounced among workers exposed to herbicides. A few chromatid exchange figures were found in the exposed persons.

Of special interest are the data on a presumably combined long-term effect described by Bartel.[57] The retrospective cohort study examined cancer morbidity in a group of men with more than 5 years occupational exposure to pesticides, beginning their work between 1948 and 1978. The ranging order corresponded to the distribution expected in the general male population, except for bronchial carcinoma, where 50 cases were observed against 27.5 expected. The dose-effect relationship was suggested by the positive correlation between the duration of employment and the lung cancer mortality. The smoking habits of the exposed men were not different from those of the general population.

Roan et al. compared some reproduction indices in 314 families of agricultural pilots and 170 control families.[58] They found no differences in the number of male and female children, spontaneous abortions, or birth defects. The authors recommended similar studies on female subjects exposed to pesticides. They recognized that the comparatively limited sizes of their study groups, was not a sufficient basis to accept a complete absence of pesticide effect on the reproductive parameters studied.

VI. CONCLUSION

After a multiple combined exposure to pesticides workers may be manifested by a more or less specific pathology and allergy.

Most often, combined occupational intoxications are subacute and observed mainly when agricultural workers do not follow the prescribed hygiene. Chronic forms are less frequent and found in workers who produce and formulate pesticides.

Chronic intoxication from combined pesticides in agriculture requires profound study.

The review of literature on chronic occupational intoxications showed a number of diagnostic difficulties, even in cases of the isolated action of a single pesticide. Only a few cases of chronic

TABLE 2
Summarized Symptoms and Syndromes at Combined Exposure to Pesticides

Nervous system	Asthenovegetative syndrome, vegetative polyneuritis, radiculitis and diencephalitis,vegetative vascular dystonia, brain sclerosis
Vision	Retrobulbar neuritis with reduced visual acuity, angiopathy of the retina, reduced corneal sensitivity, narrowed visual field
Respiratory system	Chronic tracheitis, initial pneumofibrosis, lung emphysema, bronchial asthma
Cardiovascular system	Chronic toxic myocarditis, chronic coronary insufficiency, hypertonia, hypotonia
Liver	Chronic hepatitis, cholecystitis, hepatocholecystitis, disturbance of detoxifying function and other liver functions
Kidneys	Disturbances, expressed in: albuminuria, azotemia, urea, creatinin, reduced excretory function, low clearance
Gastrointestinal tract	Chronic gastritis, duodenitis, ulcus, chronic colitis (hemorrhagic, spastic, quasipolyposic neoplasms), secretory disturbances, (hypersecretion, hypoacidity up to achylia), motor disturbances
Hemopoietic system and blood	Leukopenia, increase of reticulocytes and lymphocytes, eosinopenia, lymphopenia, monocytosis, anemia
Skin	Dermatitis and eczema

TABLE 3
Scheme for Medical Examinations

Prophylactic examinations		
Obligatory (in out patient clinic)	**Recommended (in out patient clinic)**	**In case of intoxication**
Serum ChE, serum protein fractions, SGPT, SGOT, routine blood and urine tests	Total protein, total bilirubin alkaline phosphatase, aldolase ECG, Ro-graphy of the lung	In cases of suspected poisoning, in the presence of some clinical symptoms, the tests are carried out in a hospital Serum ChE, serum protein fractions, SGPT, SGOT, routine blood and urine test, total protein, total bilirubin, alkaline phosphatase, aldolase, ECG, Ro-graphy of the lung, lactate dehydrogenase, (LDH1/LDH5 ratio, ornithyl carbamyltransferase, leucinaminopeptidase, blood lipids (beta-lipoproteins, total cholesterin), specific metabolites, EEG

intoxications of agricultural workers are reported in literature. The problem with the combined effect of pesticides is even more complicated.

Chronic intoxication is not the only possibility for the adverse effects pesticides have on man. Alterations in the structure of the general morbidity may also be an expression of the combined chronic effect of pesticides.

The difficulties in diagnosing "chronic combined intoxication with pesticides" are evident; however, such intoxication is quite realistic. Therefore research efforts should be developed in this direction.

Table 2 attempts to summarize the data already cited, as well as some from additional sources.[57-71] Such a synopsis reveals that damage of all organs and systems in the organism is possible under combined pesticide exposure.

REFERENCES

1. **Murphy, R. S., Kutz, F. W., and Strassman, S. C.,** Selected pesticides residues or metabolites in blood and urine specimen from a general population survey, *Environ. Health Perspectives,* 48, 81, 1983.
2. **Fournier, E.,** Phénomènes biochimiques observées en niveau du foie au cours des hépatites toxique, in *Compts rendus 7e réunion national des centres de lutte contre les poisons,* Paris, Nasson, 1966, 12.
3. **Tocci, P. M., Mann, J. B., Davies, J. E., and Edmunson, W. F.,** Biochemical differences found in persons chronically exposed to high levels of pesticides, *Ind. Med. Surg.,* 38, 188, 1969.
4. **Warnick, S. L. and Carter, J. E.,** Some findings in a study of workers occupationally exposed to pesticides, *Arch. Environ. Health,* 25, 265, 1972.
5. **Kaloyanova-Simeonova, F.,** *Pesticides Toxic Action and Prophylaxis,* Bulgarian Academy of Science, Sofia, 1977, (in Bulgarian).
6. **Spassovski, M.,** Experimental investigations on combined effect of some organophosphorous and organochlorine insecticides, PhD Thesis, 1962, (in Bulgarian).
7. **Dinman, B.,** Acute combined toxicity of DDVP and chlordane, *Arch. Environ. Health,* 9, 765, 1964.
8. **Steigalo, E., Givogliadov, L., and Druchevskaja, Z.,** On the combined effect of organophosphorous and organochlorine compounds on the growing organism, *Gig. Primen. Toksikol. Pestits. Klin. Otravlenii,* 498, 1968, (in Russian).
9. **Monseur, J.,** Intoxications par mélange d'insecticides chlorés et phosphorés, *Bull. Méd. Lèg. Toxicol. Méd.,* 7, 6, 450, 1964.
10. **Soubrier, R., Bresson, R., and Tolot, F.,** Manifestations pathologiques en relation avec l'emploi de pesticides en désinfection et de désinsectization, *Arch. Mal. Prof.,* 26, 6, 339, 1965.
11. **Brachfeld, J. and Zavon, J.,** Organophosphate (Phosdrin) intoxication, *Environ. Health,* 6, 11, 259, 1965.
12. **Thiodět, J., Massonnet, J., Milhaud, M., Colona, P., and Tenati, E.,** Intoxication par insecticides à base de parathion. Intéret des methodes de réanimation respiratoire dans le traitement des formes graves, Sté Médicales des Hopitaux d'Alger A. M. Février, 1960, 183.
13. **Izmirova, N., Petkova, V., and Beraha, R.,** Sanitary-hygienic investigations on mechanized application of pesticides, *Hig. Zdraveop.,* 24, 6, 537, 1981, (in Bulgarian).
14. **Lings, S.,** Pesticide lung: a pilot investigation of fruit-growers and farmers during the spraying season, *Br. J. Ind. Med.,* 39, 4, 370, 1982.
15. **Barthel, E.,** Irritating and allergic effect of aerosol pesticides on respiratory ways and the problems of expertise, *Z. Gesamte Hyg. Ihre Grenzgeb.,* 19, 11, 678, 1983, (in German).
16. **Lehnigk, B., Thiele, Ed., and Wosnitzka, H.,** Disturbances of respiratory functions due to the effect of agricultural chemicals, *Z. Erkr. Atmungsorgane,* 164, 3, 267, 1985, (in German).
17. **Anon.,** Report of an International Workshop on Epidemiological Toxicology of Pesticides Exposure, *Arch. Environ. Health,* 25, 339, 1972.
18. **Stoyanov, T. G.,** Changes in some biochemical indices at chronic effect of pesticides, *Transp. Med. Vesti,* 24, 1, 26, 1979, (in Bulgarian), (summary in German and Russian).
19. **Petkova, V. and Jordanova, Ju.,** Enzyme changes at chronic exposure to pesticides, *Probl. Hig.,* 5, 133, 1980, (in Bulgarian).
20. **Bezuglyi, V. P. and Kaskevish, L. M.,** Some indices of the functional state of liver in flight personnel engaged in aerial chemical operations, *Gig. Tr. Prof. Zabol.,* 13, 10, 52, (in Russian).
21. **Trendafilova, R., Charakchiev, D., Krasteva, S. and Kolarska, S.,** Laboratory data in subjects with occupational combined exposure to pesticides, *Savrem. Med.,* 34, 1, 21, 1983, (in Bulgarian).
22. **Zolotnikova, G. P.,** On the early detection of functional liver disturbances in women with occupational exposure to pesticides in greenhouse conditions, *Gig. Sanit.,* 2, 24, 1980 (in Russian).
23. **Izmailova, T. D. and Derevyanko, L. D.,** Hygienic characteristics of working conditions and the functional state of technics from agricultural aviation, *Gig. Tr. Prof. Zabol.,* 9, 27, 1986, (in Russian).
24. **Bezuglyi, V. and Komarova, L. M.,** On the syndrome of Gilbert in subjects exposed to pesticides, *Vrach. Delo,* 3, 105, 1984, (in Russian).
25. **Kaskevich, L. M. and Djogan, L. K.,** The state of some metabolic processes in subjects exposed to pesticides, *Gig. Tr. Prof. Zabol.,* 10, 47, 1983, (in Russian).
26. **Morgan, D. and Roan, C.,** Renal function in persons occupationally exposed to pesticides, *Arch. Environ. Health,* 19, 5, 633, 1969.
27. **Gombert, A., Beat, V., Bonderman, P., and Long, K.,** Seasonal pesticide exposure relationships including renal phosphate reabsorption and urinary alpha-aminoacid nitro loss, *Arch. Environ. Health,* 21, 2, 128, 1970.
28. **Zaykov, C.,** Immunotoxicology of pesticides, in Toxicology of Pesticides, Proc. of a Seminar, Sofia, Bulgaria, 1981, World Health Organization Regional Office for Europe, Copenhagen, 78, 1982.
29. **Gervais, P.,** Les accidents allergiques en agriculture, *La Révue Praticien,* 25, 3841, 1965.
30. **Gervais, P.,** Reactions allergiques aux substances chimique de composition définie, Physiopathologie, Clinique, Diagnostique, Masson et Cie, 1968, 350.

31. **Vacher, J. and Vallet, G.,** Les allergie supposées aux pesticides et engrais. Dix années d'observations partielles en France, *Arch. Mal. Prof.,* 29, 6, 336, 1968.

32. **Kambo, Y., Matsushima, S., Matsumura, T., Kuroumo, K., and Suzuki, N.,** Studies on patch tests in contact dermatitis caused by pesticides, *Whither Rural Medicine,* Proc. 4th Int. Congr. Rural Med., Usuda, 1969, (Tokyo), 1970, 55.

33. **Kozintseva, P., Kuznetsova, L., and Zinchenko, D.,** Clinical significance of the detection of antipesticide circulating antibodies, *Gig. Primen. Tokik. Pestits. Klin. Otravlenii,* 345, 1973, (in Russian).

34. **Popov, I., Darlenski, B., and Mladenova, S.,** Latent allergy to pesticides in the developing of occupational contact dermatites in greenhouse workers, *Dermatol. Venerol.,* 23, 2, 21, 1984, (in Bulgarian).

35. **Karimov, A. M.,** Occupational skin diseases in cotton growers caused by chemical poisons and measures for their prevention, *Gig. Tr. Prof. Zabol.,* 14, 35, 1970, (in Russian).

36. **Katsenovich, L. A., Rusybakiev, R. M., and Fedorina, L. A.,** T- and B-systems if immunity in patients with pesticide intoxication, *Gig. Tr. Prof. Zabol.,* 4, 17, 1981, (in Russian).

37. **Pol'chenko, V. I. and Zinchenko, D. V.,** Immunity indices in epidemiological studies of children from districts with intensive and limited use of pesticides, in *Methodology, Organization and Some Mass Immunological Examinations, Summaries of the Reports of an All-Union Conference, Angarsk 1987,* Petrov, R. V., (Ed.), Moskva, Angarsk, 1987.

38. **Zolotnikova, G. P.,** Disturbances of the immunological reactivity of organism under exposure to pesticides in greenhouse conditions, *Gig. Tr. Prof. Zabol.,* 3, 38, 1980, (in Russian).

39. **Bezuglyi, V. P.,** The effect of pesticides and their complexes on the development of non-specific diseases, 7, 102, 1980, (in Russian).

40. **Chezzo, F., Berretti, P., Perini, G., Belloni, G., and Vigan, L.,** Long-term effects of pesticides on human health, epidemiological research on the morbidity of rural population, *Minerva Med.,* 59, 2949, 1968.

41. **Janicki, K.,** Leukemias in the Krakow region in the years 1961–1968 and contamination of the environment by pesticides, *Acta Med. Pol.,* 13, 1, 49, 1972.

42. **Romash, A. V. and Dorofeev, V. M.,** Significance of some indices of peripheral blood in the diagnosis of chronic intoxications by pesticides, *Lab. Delo,* 11, 679, 1984, (in Russian).

43. **Sutneeva, G. I.,** Sanitary-hygienic conditions of application and toxicological properties of the mixture of keltan, chlorofos and copper hydroxide, *Gig. Sanit.,* 8, 101, 1973 (in Russian).

44. **Tsvetkova, T., Atanassova, K., and Andonov, S.,** Cation proteins content of granulocytes in agricultural workers, *Hig. Zdraveop.,* 27, 3, 216, 1984, (in Bulgarian).

45. **Long, K. L., Beat, V. B., Gombart, A. K., Sheeta, R. F., Hamilton, H. E., Falaballa, F., Bonderman, D. P., and Choi, U. Y.,** The epidemiology of pesticides in a rural area, *Am. Ind. Hyg. Assoc. J.,* 30, 298, 1969.

46. **Fokina, K.,** Rheoencephalographic diagnosis of vascular pathology of the brain in persons exposed to a complex of pesticides, *Gig. Tr. Prof. Zabol.,* 4, 49, 1980, (in Russian).

47. **Fokina, K. V.,** Nervous system disturbances in persons working with cuprious and other pesticides, *Vrach. Delo,* 9, 113, 1984, (in Russian).

48. **Röder, H. and Lindher, M.,** Electroneurographic studies of agricultural workers exposed to pesticides, *Ergeb. Exp. Med.,* 41, 379, 1982, (in German).

49. **Kalič-Filipovič, D., Dodič, S., Savič, S., Prodanovič, M., Arsenijevič, M., Guconič, M., and Kidakovič, A.,** Occupational health hazards in the production and application of pesticides, *Arch. Hyg. Rada,* 24, 3, 333, 1973.

50. **Nuritdinova, F.,** The effect of pesticides applied in agriculture upon the organ of vision, *Med. Zdravoohr. Uzb.,* 6, (in Russian).

51. **Glazko, I.,** The frequency of eye impairments and visual functions in persons working with toxic chemicals in agriculture of Donets region, *Gig. Tr. Prof. Zabol.,* 14, 1, 34, 1970 (in Russian).

52. **Campbell, A.,** Neurological complications associated with insecticides and fungicides, *Brit. Med. J.,* 2, 415, 1952.

53. **Imazumi, K., Atzumi, K., Horie, E., Mita, K., Takano, Y., Goto, I., Nakumura, R., and Kashii, K.,** Ophthalmic disturbances due to agricultural chemicals in Japan. *Whither Rural Medicine,* in Proc. 4th Int. Congr. Rural Med., Usuda, 1968 (Tokyo), 1970, 47.

54. **Sugaya, R.,** Studies on pesticide poisoning among farmers, in *Whither Rural Medicine,* in Proc. 4th Int. Congr. Rural Med., Usuda, 1968 (Tokyo), 1970, 47.

55. **Trusiewiczowa, D.,** Pesticides and organ of vision, *Klin. Oczna.,* 85, 6, 263, 1983.

56. **Joder, J., Watson, M., and Benson, W.,** Lymphocyte chromosome analysis of agricultural workers during extensive occupational exposure to pesticides, *Mutat. Res.,* 21, 335, 1973.

57. **Barthel, E.,** Increased risk of lung cancer in pesticide-exposed male agricultural workers, *J. Toxicol. Environ. Health,* 8, 1027, 1981.

58. **Roan, C. C., Matanoski, G. E., McIlnay, C. Q., Olds, K. L., Pylant, F., Trout, J. B., Wheeler, P., and Morgan, D. P.,** Spontaneous abortions, stillbirths and birth defects in families of agricultural pilots, *Arch. Environ. Health,* 39, 1, 56, 1984.

59. **Engberg, I., De Bruin, A., and Zielhuis, R.,** Health of workers exposed to a cocktail of pesticides, *Int. Arch. Arbeitsmed.,* 32, 171, 1974.

60. **Alexeeva, G. and Jakubova, R.,** The effect of pesticides on the health of population under conditions of their application in Uzbekistan, *Gig. Primen. Toksikol. Pestits. Klin. Otravlenii.,* 455, 1969, (in Russian).

61. **Bezuglyi, V., Muhtarova, N., Buslenko, A., Kachalaj, D., Komarova, L., and Bereznyakov, I.,** The health state of persons with long-term occupational exposure to pesticides, *Gig. Primen. Toksikol. Pestits. Klin. Otravlenii,* 417, 1967, (in Russian).

62. **Bezuglyi, V., Muhtarova, N., Kaskevich, L., Kachlaj, D., Komarova, L., and Bereznyakov, I.,** On the health state of persons from the flight-technical personnel engaged with aviation chemical work, *Gig. Primen. Toksikol. Pestits. Klin. Otravlenii,* 466, 1968, (in Russian).

63. **Barseva, L.,** On the chronic effect of plant-protection chemicals and the functional state of liver in people working with them, *Hig. Zdrav.,* 9, 5, 491, 1966, (in Bulgarian).

64. **Kagan, Y., Trachtenberg, I., Rodionov, G., Lukaneva, A., Voronina, L., and Verich, G.,** The effect of pesticides upon the rise and development of some pathological processes, *Gig. Sanit.,* 3, 23, 1974, (in Russian).

65. **Krivoglaz, B.,** Clinic and treatment of poisonings due to toxic chemicals, *L.,* 1965, 165, (in Russian).

66. **Sharp, D. S., Eskenazi, B., Harrison, R., Calas, P., and Smith, A. H.,** Delayed health hazards of pesticide exposure, *Ann. Rev. Public Health,* 7, 441, 1986.

67. **Kundiev, Y., I., Krasnyuk, E. M.,** Eds., Occupational diseases in workers engaged in agriculture, Kiev, Zdorovje, 1983 (in Russian).

68. **Hayes, W.,** Pesticides and human toxicity, *Ann. N.Y. Acad. Sci.,* 160, 40, 1969.

69. **Kachalaj, D.,** Functional state of the gastrointestinal tract of warehousemen in stores for toxic chemicals, *Gig. Primen. Toksikol. Pestits. Klin. Otravlenii,* (Kiev), 1968, 253, (in Russian).

70. WHO Technical Reports Series, No 571, Early detection of health impairment in occupational exposure to health hazards, World Health Organization, Geneva, 1975.

71. **Medved, L. I., Kundiev, Y. I.,** Hygiene of labour in agriculture, *M. Med.,* 1981, (in Russian).

Chapter 14

EPIDEMIOLOGY OF THE ACUTE PESTICIDE INTOXICATION

The available information on the epidemiology of acute pesticide poisonings is very fragmentary. Getting a clear idea of the seriousness of the problem, in the world as a whole is not possible; in some regions information is missing, because the intoxications are not registered, or not even known. Nevertheless, the WHO working group and some other authors attempt to estimate the possible number of intoxications.

Based on statistical data for 19 countries the WHO working group estimated that, worldwide, there have been as many as 500,000 cases annually of pesticide intoxications. The mortality rate has been 1% in countries where medical treatment is readily available, so deaths are estimated to be about 5,000 a year.[1]

However, it is reckoned that the real number of poisonings is considerably greater — about 2 million per annum. About 40,000 persons die and admittedly 75% of the lethal cases occur in developing countries. That means that every 17 min a person dies of pesticide poisoning, and every 10 min 28 subjects are poisoned by pesticides.[2]

The WHO working group recently estimated pesticide intoxications based on the population at risk and the rate of intoxication.[3] Suicide attempts were estimated to be about 2 million, occupational intoxications — 700,000, and other unintentional cases — 300,000. Mass poisoning outbreaks from contaminated food were estimated to be about 1,650. Some data on acute intoxications in different countries are given in Table 1.

In past years (1956–1967), many mass poisonings were due mainly to the contamination of food products, together with the use of highly toxic pesticides, such as endrin, parathion, and hexachlorbenzene.[11] The mortality rate in some of these incidents was more than 10%. After 1970 they still occurred, but for different reasons (Table 2).

The real morbidity and mortality rate of pesticide intoxications is difficult to calculate in view of the poor registration and the size of the exposed population. In the publications available, the annual incidence rate varies from 6 to 79 per 100,000 agricultural (risk) population.

Levine reviewed the available data on unintentional pesticide poisonings.[19] The majority were likely due to occupational exposure in agriculture. The annual rates of unintentional pesticide poisonings range from 0.3 to 18 cases per 100,000 in 17 countries.

Mortality rate in the U.S. has been given as 1 per 1 million; the ratio between fatal and nonfatal poisonings was 1:13 in one study and 1:75 in another, depending on the severity (Hayes, 1964, cited by Mrak[20]). Pesticide intoxications represent 6 to 12.8% of the total number of intoxications by solids and liquids. Parathion was the leading cause of death (49%) in the groups under 5 years of age (McCarthy, 1967, cited by Mrak[20]).

In California the number of persons at risk is approximately 250,000. Annually 827 to 1013 cases of occupational injury from pesticides were reported between the years 1960 and 1963 (West and Milby, 1965, cited by Mrak[20]).

Caldwall and Sundifer reported that the cases of pesticide poisonings in South Carolina were 20 per 100,000.[21]

Hayes and Vaughan studied mortality from pesticides in U.S. for 1973 and 1974.[22] Most of the poisonings were nonoccupational and involved gross carelessness. Before 1961 the number of deaths was more than 100 per year, while in 1973 it was 32 and in 1974 it was 35.

Morgan et al. studied morbidity and mortality in workers occupationally exposed to pesticides.[23] Proportionate mortality patterns were similar in exposed and control groups. Skin diseases and skin cancer were exceptionally common among pest control operators. Elevated serum DDT and DDE levels appeared to characterize persons who reported hypertension, atherosclerosis, and possibly diabetes in subsequent years.

TABLE 1
Acute Pesticide Poisonings

Country	Period	Number	Kind of pesticide	Type of injury	Reasons for poisonings	Authors
Bulgaria	1954–77	1559 occupational	Above 50% OP	Acute intoxications	Work in treated areas	Kaloyanova[4]
Costa Rica	1985–86	236 (160 occupational)				Taylor et al.[5]
Egypt	1971–76	3000 unspecified	Leptophos, tamaron and gusathion			Sallam an El Chawaby[6]
	1979	4000 30 fatal occupational	Lanate			
England and Wales	1952–71	296 9 fatal (3 occupational)			Handling concentrates and containers Disposal of empty containers Inadequately labeled containers	Hearn[7]
France	3 years	Occupational	OP – 47 Nitrophenols – 36 Arsenicals – 24 Mercury compounds – 22 Methylbromide – 15 Others – 27			Dubrisay and Fags[8]
Guatemala	1981	787 1 fatal	OP – 49.5 Carbamates – 15.50 Organochlorines – 12 Gramoxone – 3.5			Taylor et al.[5]
	1985	786 15 fatal				
Hungary	1971–76	803 occupational 11 fatal (fosdrin - 5, other OP - 2)	OP – 389 OC – 9 Carbamate – 20 Nicotine – 7 Herbicide – 221 Fungicides – 94 Rodenticides – 4 Fumigants – 9 Others – 50			Adamis and Szenzenstein[9]

Country	Period	Cases	Agents	Effects	Causes	Reference
Japan	1970–77	1376 Clinical cases (including nonoccupational) 728 men 648 women 87 fatal	OP – 30% OC – 6% Carbamates – 5% Herbicides – 12.5% Organosulphurus – 10% Organomercurial – 1% Antibiotics – 3% Nicotine – 66	Acute intoxication – 46% Eye injuries – 0.9% Nasolaryngopharying injuries – 42.6% Others – 5.6%	Insufficient protection – 40% Poor general health – 19% Improper weather conditions – 5%	Sugaya and Wakatsuki[10]
	12 years	204 occupational (apple orchards)				Matsumitsu et al.[12]
	1978	167 clinical cases 87 men 78 women 113 occupational 15 fatal 14 suicides 11 paraquat	OP – 63 Herbicides – 20 Sulphur – 19 Carbamate – 4 Antibiotics – 5 Others – 48	Acute intoxication – 99 Skin impairment – 54 Ocular impairment – 3	Insufficient protection – 40% Poor general health – 19% Improper weather conditions – 5%	Sugaya and Wakatsuki[11]
Malaysia	Annually	1200 estimated 9 fatal				Jearatnam et al.1[13]
Mexico		847 occupational 4 fatal	Methylparathion – 16.5% Sevin – 11% Azodrin – 8.9% Lannate – 6.6% Gusathion Methyl – 2.3% Other OP – 14.8% OC – 1.6%			Nagera and Sánchez de la Fuente[14]
Nicaragua	1984	395 5 fatal				Taylor et al.[5]
Sri Lanka	1975–80	Annually 13,000 hospitalized 1000 fatal			73.1% suicides 17.1 occupational	Jeyaratnam et al.[13]
California	1974	1257 occupational				Durham[15]

TABLE 2
Mass Poisonings After 1970

Country	Number	Pesticide	Authors
Iraq	6000 (500 fatal)	Methylmercury in food, bread, etc.	Bakir et al.[16]
Pakistan	2800 (5 fatal)	Malathion (occupational exposure)	Bakir et al.[17]
U.S.	1350 (8 fatal)	Aldicarb (water, melon contamination)	Green et al.[18]

On the basis of 27 pesticide intoxications in 436 citrus growers, Griffith and Duncan estimated 376 citrus fieldworker related poisonings per year in Florida.[24] The incidence rate was 113 poisonings per 10,000 fieldworkers; 75 were hospitalized. That is an incidence rate of 23 hospitalized cases per year for every 10,000 field workers. The estimated incidence rate for possible poisonings in citrus field workers was 295 per 10,000. A later estimate came up with 34 confirmed and 160 possible poisonings per 10,000 field workers.[25]

In a Philippine rice growing area the mortality rate, related to diagnosed pesticide poisoning in men aged 15 to 54, increased from 2.39 to 8.29 per 100,000 for the periods 1961 to 1971 and 1972 to 1984, respectively. The mortality rate of leukemia also increased, while the overall cancer rates have not increased significantly.[26] The author discussed a possible link between exposure and the development of leukemia.

Jeyaratnam et al. surveyed acute pesticide poisonings among agricultural workers in Indonesia, Malaysia, Sri Lanka, and Thailand.[27] Pesticide users among agricultural workers ranged from 29.8% in Indonesia to 91.9% in Malaysia. About 10 to 20% of the pesticide users suffered an episode of pesticide poisoning during their working life. Among hospital admissions in 1983 (28 to 107 cases), unintentional occupational and nonoccupational poisoning was responsible for 32% in Malaysia, 62% in Sri Lanka, 22.7% in Thailand, and 2% in Indonesia. The percentage of suicides varied from 36.2 in Sri Lanka to 67.9 in Malaysia. The annual incidence rate of pesticide poisoning in Sri Lanka for the period 1975 to 1980 was 79 per 100,000. The equivalent rate for suicide attempts was 58 per 100,000, and for occupational intoxication it was 13 per 100,000.[7] The fatality rate was about 9%.[7,27]

Acute pesticide intoxications are a high priority in developing countries. We cannot agree that these intoxications are "third world diseases" as stated by Agarwal (cited by Sim[28]), because this problem exists in all countries, although at a different level. Thus, in Sri Lanka two and a half times as many people were hospitalized as in the U.S., and nearly five times as many died, despite of the fact that the U.S. uses about a third part of the whole pesticide in the world. Three pesticide poisoning admissions into Jakarta hospitals take place each day, as well as in one large hospital in Bangkok.

In Sri Lanka, 50% more pesticide related deaths occured in 1977 than deaths from malaria, tetanus, diptheria, whooping cough, and polio taken together.[28]

Epidemiological study of chronic and long-term effects is a difficult task. Many confounding factors make it almost impossible to establish a clear cause-effect relationship. Usually, the pesticides contribute to a quicker emergence of some disease already existing in a latent stage, or exacerbate some illness.

The prevalence of some diseases in the mortality of two groups of different pesticide exposure may indicate such an effect. The rate of very mild subclinical intoxications would be much higher. Some indication for this suggestion comes from the biological monitoring of ChEA, performed regularly in some countries. In selected groups of farmers using pesticides, those with inhibited ChEA reach up to 30% in some countries.

About 24% of 821 pesticide users in Malaysia had ChE inhibition by more than 25 to 30%.[29] In Sri Lanka, 15% of the workers in a pesticide formulating factory and 25 to 30% of the workers in agriculture have manifested unsatisfactory ChEA levels. In Brazil, ChEA inhibition has been found in about 20% of the exposed workers.[30]

In Bulgaria, biological monitoring of ChEA is performed on all persons engaged in pesticide spraying in organized teams. The data are given in Table 3.[31]

In a study of self-reported pesticide poisonings and hospital admissions for poisoning, about 7% of the agricultural workers using pesticides in Malaysia and Sri Lanka had been poisoned in the course of 1 year.[27] Organophosphorous compounds were responsible for the majority of acute intoxications and lethal cases.

Polchenko analyzed 34,000 intoxications in 66 countries from 1945 to 1966 and came to the following principal conclusions: there was no correlation between the number of intoxications and the amount of pesticides used; 75.4% of the intoxications were produced by organophosphorous pesticides (80% by parathion); 66.6% of the intoxications were occupational; 23% of them occurred during pesticide production and 77% during application (about 20% of the latter were reentry intoxications).[32]

An analysis of poisoning cases in Bulgaria from 1965 to 1968 showed that 28% were of persons engaged in treated areas; for the period of 1968 to 1973 this percentage increased to 56%, due to the increased aircraft application.[11] Of the total number of intoxications, 69.5% were provoked by organophosphates.

Kahn reckons that officially reported cases are only a small fraction of the actual number of such poisonings.[33] In Dade County, Florida, 72 pesticide fatalities (46%) between 1959 and 1965 were due to parathion.[20] In a study of 129 poisonings in South Texas, 98% were caused by parathion and methylparathion.

In the U.S., parathion was also the leading cause of death (49%) for children under 5.[20]

Paraquat is often responsible for intoxications — mostly suicidal — in Malaysia, as well. From 1977 to 1981 post mortem analysis of poisoning cases showed the most commonly consumed poisons: herbicides — 326 cases (310 cases or 95% with paraquat), OP — 155 cases (85% malathion and 4% dichlorovos), and OCP — 88 cases (31.8% DDT, 26% thiodan, 18% dieldrin, and 15% BHC). Of the 565 total deaths, 54.5% were due to paraquat. In one state of Malaysia the prevalence of paraquat poisonings per 100,000 population was calculated to be 4.6% (73.4 suicides: 13.8% accidental and 1.7% occupational).[28]

In Sri Lanka and Malaysia, the major class of identified chemicals causing pesticide poisoning are organophosphates, 65% and 53.6%, respectively. In Thailand, bipyridiliums are responsible for 25% of the intoxications, and in Indonesia, copper compounds are responsible for 23.4%.[27]

The reasons for poisonings are generally the same in all countries: inadequate technical equipment, insufficient technological training, poor personal hygiene and health education, nonuse of educational protective devices, inadequate labeling and disposal of containers, use of highly toxic substances, increased working days, and work in treated areas. The "pesticide residue related illness" or "reentry poisonings" are more frequently reported.

Polchenko reviewed the world literature on acute pesticide intoxications.[34] Polchenko analyzed 70,000 cases of intoxication and proposed a 9 criteria classification. With some modifications, this classification is shown in Table 4.

Polchenko only analyzed the nonoccupational unintentional intoxications, published in the literature.[35,36] About 47% of these intoxications are in the 0 to 5 age group. The poisonings involve the use and storage of pesticides at home (12.3%), the consumption of contaminated food, or contact with pesticide residues in the environment (87.7%).

CONCLUSION

Information about the epidemiology of acute pesticide intoxications is inadequate. Nevertheless, it is obvious that the problem is serious, particularly in countries without well developed and implemented preventive strategies for pesticides use.

TABLE 3
ChEA in Monitored Workers

Year	Before working season			During spray season			End of the working season		
	Total	Inhibited	%	Total	Inhibited	%	Total	Inhibited	%
1975	21,294	1,947	9.14	8,033	589	7.33	1,176	269	22.87
1976	26,761	2,894	10.80	6,484	322	4.90	1,769	145	8.19
1977	92,982	1,345	5.80	9,345	405	3.30	2,019	153	7.47
1978	31,376	2,626	6.33	8,618	556	6.45	2,494	267	10.70
1979	27,522	1,911	6.90	10,079	388	3.84	2,227	127	5.70
1980	28,941	3,360	11.60	10,758	166	1.45	2,237	147	6.57
1981	18,011	1,964	10.90	8,458	992	11.70	2,869	140	4.87

TABLE 4
Classification of Intoxicaitons by Pesticides[34]

Criteria	Classes
Character	Accidental
	Occupational
	Industry
	Agriculture (including reentry)
	Public health
	Forestry
	Trade and other
	Nonoccupational
	Home
	Environment
	Primary
	Secondary (via environmental media)
	Intentional
	Suicide
	Homocide (criminal intoxications)
Route of entry	Oral
	Respiratory
	Dermal
Number of intoxicated persons	Single – up to 5 persons
	Group poisoning – 5 to 100 persons
	Mass poisoning – more than 100 persons
Grade of intoxication	Severe
	Lethal
	Nonlethal
	Medium
	Light

The WHO/UNEP Working Group held in Tbilisi in 1987, estimated that 3 million pesticide intoxications occur annually all over the world. Suicide should be responsible for 2 million cases and occupational exposure for 700,000; the remaining 300,000 are due to other unintentional reasons. Organophosphorous pesticides, paraquat, and carbamates most often are responsible for intoxications.

The mortality rate of pesticide poisonings varies from 1 to 83 cases per million relative to the development of the country and its economic structure. The incidence rate may reach 79 per 100,000.

Reentry intoxication is a problem recognized in some countries. It was estimated that the incidence rate of such intoxications may be about 160 cases per 10,000 workers. This kind of intoxication is mainly related to the use of organophosphorous compounds, but also to carbamates and pesticides from other groups.

REFERENCES

1. Safe use of pesticides, WHO Expert Committee on Insecticides, 20th Report, TRS No. 513, World Health Organization, Geneva, 1973.
2. Ecoforum, Ind. Toxic. Res. Centre, Lucknow, *Ind. Toxicol. Bull.,* 6, 4, 19, 1982.
3. Public health impact of pesticides used in agriculture, Report of WHO/UNEP working group, May, 1988, World Health Organization, Geneva.
4. **Kaloyanova-Simeonova, F.,** *Pesticides. Toxic Action and Prophylaxis,* Publishing House of the Bulgarian Academy of Science, Sofia, 1977.
5. **Taylor, R., Tame, K., and Goldstein, G.,** Paraquat poisoning in Pacific Island countries 1975–85, South Pacif. Comm. Noumea New Caledonia, Dec. 1985, T. Paper 18.

6. **Sallam, M., El-Ghawaby, S. H.,** Safety in the use of pesticides, *J. Environ. Sci. Health, B.,* 15, 6, 677, 1980.
7. **Hearn, C. E. D.,** A review of agricultural pesticide incidents in man in England and Wales, 1952, 1971, *Br. J. Ind. Med.,* 30, 3, 253, 1973.
8. **Dubrisay, J. and Fages, J.,** Occupational pathology in agricultural activities. Attempts of a statistical approach. II. Occupational diseases., *Arch. Mal. Prof.,* 39, 7, 459, 1978.
9. **Adamis, Z. and Szenzenstein, M.,** Occupational poisonings in agriculture, in Konferenz über Sicherheitstechnik der landwissen-schaftlichen Chemisierung. Vorträge, OMKDK Technoinform, Budapest, 1978, 9.
10. **Sugaya, H. and Wakatsuki, S.,** Countrywide investigation of clinical cases of pesticide intoxication (collected during 1970–1977), *Jpn. J. Rural Med. (Nippon Nosen Igakkai Zasshi),* 27, 3, 436, 1978.
11. **Sugaya, H. and Wakatsuki, S.,** Results of a nationwide survey on clinical intoxication cases due to agricultural chemicals, *Jpn. J. Rural Med.,* 28, 3, 452, 1979.
12. **Masumitsu, T., Nagata, J., Kobayashi, H., and Ono, S.,** Studies on intoxication due to a natural insecticide, nicotine, (Report I) *Nippon Nosen Igakkai Zasshi,* 29, 1, 61, 1980.
13. **Jeyaratnam, J., de Alwis, R. S., and Copplestone, J. P.,** Survey of pesticide poisoning in Sri Lanka, *Bull. of WHO,* 60 (4), 1982.
14. **Nagera, R. R. and Sanchez de la Fuente, E.,** Pesticide intoxications in "Comarca Lagunera" during 1974, *Salud Publica Mex.,* 17, 5, 687, 1975.
15. **Durham, W. P.,** Pesticides and human health, in *Pesticides. Contemporary Role in Agriculture, Health and the Environment,* Sheets, T. J. and Pimentel, D., Eds.. Humana Press, Clifton, NJ, 1979, 83.
16. **Bakir, F., Damluji, L. S., Amin-Zaki, L., Murtadha, M., Khalidi, A., Al Roni, N., Tikritis, Dhahir, H., Clarkson, T., Smith, J., and Doherty, R.,** Methylmercury poisonings in Iraq, *Science,* 184, 230, 1973.
17. **Baker, E. L., Zack, M., Miles, J. W., Alderman, L., Dobbin, R. D., Miller, S., Teeters, W. R., and Warren, M.,** Epidemic malathion poisoning in Pakistan malaria workers, *Lancet,* 1, 8054, 31, 1978.
18. **Green, M. A., Neumann, M. A., Wehr, H. M., Foster, L. R., Williams, L. P., Polder, J. A., Morgan, C. L., Wagner, S. L., Wanks, L. A., and Witts, J. M.,** An outbreak of watermelon born pesticide toxicity, *Am. J. Publ. Health,* 77, 1431, 1987.
19. **Levine, R. S.,** Assessment of mortality and morbidity due to unintentional pesticide poisonings, WHO/VBC, 86, 929, World Health Organization, Geneva, 1986.
20. **Mrak, E. M.,** Report of the Secretary, Commission on Pesticides and their Relationship to Environmental Health, U.S. Department of Health, Education and Welfare, 1969.
21. **Caldwell, S. T. and Sandifer, S. H.,** South Carolina pesticide poisonings, 1978, *J. S. C. Med. Assoc.,* 75, 12, 615, 1979.
22. **Hayes, W. J. and Vaughan, W. K.,** Mortality from pesticides in the United States in 1973 and 1974, *Toxicol. Appl. Pharmacol.,* 42, 235, 1977.
23. **Morgan, D. P., Lin, L. J., and Saikaly, H. H.,** Morbidity and mortality in workers occupationally exposed to pesticides, *Arch. Environ. Contam. Toxicol.,* 9, 3, 349, 1980.
24. **Griffith, J. and Duncan, R. C.,** Grower reported pesticide poisonings among Florida citrus fieldworkers, *J. Environ. Sci. Health,* 20, 1, 61, 1985.
25. **Griffith, J., Duncan, R. C., and Konefal, J.,** Pesticide poisonings reported by Florida citrus fieldworkers, *J. Environ. Sci. Health,* 20, 6, 701, 1985.
26. **Loevinsohn, M. E.,** Insecticide use and increased morbidity in rural central Luzon, Phillipines, Epidemiology, *Lancet,* June 13, 1359, 1987.
27. **Jeyaratnam, J., Lun, K., and Phoon, W. O.,** Survey on acute pesticide poisoning among agricultural workers in four Asian countries, *Bull. of WHO,* 65, 4, 521, 1987.
28. **Sim, F. G.,** The pesticide poisoning report. A survey of some Asian countries, International Organization of Consumer Unions, Regional Office for Asia and the Pacific, P.O. Box 1045, Penang, Malaysia, 1985.
29. **Jeyaratnam, J., Lun, K. C., and Phoon, W. O.,** Blood cholinesterase levels in agricultural workers in four Asian countries, *Toxicology Letters,* 33, 193, 1986.
30. **Trape, A. Z., Garcia, E. G., Borges, L. A., De Almeida Prado, M. T., Kavero, M., and Almeida, W. F.,** Trojeto de vigilancia epidemiologia om ecotoxicologia de pesticidas. Abordagem Preliminar Revista Bras. de Sal. Occup., 47, 12, 1984.
31. **Kaloyanova, F. P.,** Occupational hazards of pesticides and their epidemiology, Proc. Int. Conf. Environ. Haz., Agrochem., Alexandria, Egypt, Nov. 8–12, 1983, 1, 312.
32. **Polchenko, V. I.,** Pesticide poisonings according to data from world literature, *Gig. Primen. Toksikol. Pestits. Klin. Otravlenii,* 1968. (In Russian).
33. **Kahn, E.,** Epidemiology of field reentry poisoning, *J. Environ. Pathol. Toxicol.,* 4, 5, 323, 1980.
34. **Polchenko, V. I.,** Classification of pesticides, *Vrach. Delo,* 2, 151, 1972, (in Russian).
35. **Polchenko, V. I.,** Household pesticide poisonings abroad, *Gig. Primen. Toksikol. Pestits. Klin. Otravlenii,* 462, 1962, (In Russian).
36. **Polchenko, V. I.,** Household cases of pesticide poisoning, *Gig. Sanit.,* 10, 90, 1969. (In Russian).

Chapter 15

REENTRY PERIODS

I. DEFINITION OF PROBLEM

Workers may suffer significant exposure when they enter previously treated areas for further cultivation and hand-harvesting.[1-3]

The degree of exposure depends on many factors, including the physical properties of the pesticide and its biodegradability, the crop, the nature of the application, and the climatic conditions.

In Bulgaria reentry poisonings occur in groups working on tobacco plantations up to 15 d after spraying or in persons who pick flowers sprayed with organophosphate compounds. It has turned out that the 5 d as reentry time, determined for all pesticides, did not guarantee safety.[1,4]

Data given by U.S. authors point out that this problem exists, especially in California and Florida.

Spears et al. described poisonings from pesticide residues while picking fruits treated with parathion, paraoxon, etion, phosalone, azinphosmethyl, diocation, dioxan, and monooxan.[5]

McClure reported an incident of poisoning during grape picking.[6] From a total of 120 persons, 108 received symptoms after 2 weeks work in vineyards sprayed with phosdimethoate, methomyl, and zolon. The pollution of the leaves reached 57 ppm with zolon and 2.3 ppm with zolonoxon.

Peoples and Maddy reported a group poisoning in California involving 118 of 120 grape pickers.[7] The workers had been allowed to enter the field before the required 30 d waiting period for dialifor and phosphalone. Excessive skin exposure to the chemical resulted in intoxication. Plasma ChEA was inhibited by more than 60%, and 85 of the patients were admitted to a hospital.

In studies performed during the picking of pears sprayed with zolon, erythrocyte ChEA decreased insignificantly at exposure level 4 mg/h; therefore, this exposure was estimated to be an insignificant risk.[8]

The basic exposure in reentry intoxications occurs via skin. The formation of more toxic metabolites or interaction with other substances in the environment may take place. The pesticide residues may be more hazardous a few days after application than at the time of application.

Kundiev et al. reported poisonings in sugar beet plantations after polychlorpinen treatment.[9] After spraying polychlorpinen or lindane after spraying polychlorpinen of lindane highly toxic compounds, namely phosgen, hydrogen cyanide, chlorocyanide and chlorocyanic compounds, were eliminated up to the 75th day when, besides polychlorpinen, ammonium nitrate was used in the soil for fertilizing.

Klissenko et al. suggested that in this case the highly toxic volatile compounds had access to the air because latches of ammonia liquor with admixtures of iron carbonyl were used.[10] Iron carbonyl possibly caused the acute poisoning in humans working in the sugar beet fields. Besides ammonia and carbon oxide, other compounds were possibly formed.

Later, it was proved that some other organochlorines, in the presence of nitrate fertilizers, also eliminate poisonous gases.

Upon welding residual organochlorine pesticides (hexachloran and sodium trichloracetate) and fertilizers in sugar beet plantation soil emit toxic gases (hydrogen chloride and phosgen). These emissions start a week later and last 18 to 20 d after pesticide application. With annual treatment the emissions start a week later and last for about 2 months. Maximum HCl emissions occur at higher atmospheric temperatures (during noon hours); that of phosgen occurs during the hours of maximum UV irradiation.[11]

Interactions between pesticides and fertilizers resulting in toxic gas emissions have to be further investigated.

The hazard of methyl bromide fumigation for soil disinfection has been described by Van den Oever et al.[12] Workers cultivating soil 9 d after application may be exposed to 15 ppm.; 11 d after application no methyl bromide is detected in the air.

II. METHODOLOGY FOR ESTABLISHMENT OF REENTRY INTERVALS

Methodologies for the determination of safe reentry intervals have been developed in Bulgaria, the U.S., and the U.S.S.R.[13-28] A unified field model for reentry intervals in the U.S. has been proposed.[19-20]

The report of the working group on occupational exposure to pesticides details this problem in the U.S. This document discusses the problem of recognizing reentry-related pesticide illnesses, as well as the methodology for defining reentry intervals and necessary legislation.[29]

Nigg underlines the need for a systematic approach to the problem as well.[30] He gives preference to simple but accurate investigation methods.

Many variables should be taken into consideration in determining and evaluating reentry periods.[31,32]

Nigg and Stampes consider four principal factors: (1) dusty work environment (in Florida clay it is defined as particles of 2 μm but in California about 29 to 57% particulates are from 2 to 50 μm and they settle better on foliage); (2) toxicity of pesticide; (3) transformation to more toxic compounds (e.g., oxon); (4) environmental conditions (more oxon metabolites are found in dry climate conditions).[21]

The authors compared the findings in California and Florida and explained the great difference in recommended reentry intervals.

Kahn worked out directives to perform field studies aimed at determining safe reentry intervals for organophosphorous compounds.[25] The pesticide quantity on foliage is measured in $\mu g/cm^3$. It has been accepted that 20% ChEA inhibition in the plasma and 15% in erythrocytes, for 2 preliminary investigations indicated some effect. That is why such subjects were not admitted to the study. In the process of investigation, single subjects with more than 40% ChEA inhibition in plasma and more than 30% in erythrocytes, as compared with the initial levels, were removed; inhibitions of 20% and 15%, respectively, were considered reliable.

Spear et al. found a correlation between foliage paraoxon content and ChEA inhibition.[33] Some U.S. authors swing toward reentry determination without investigating humans. They use the relationship between pesticides on foliage and dermal LD_{50} for rats.[34,35]

The OP disappearance rate is a prime consideration for determining safe reentry times. Nigg and Stamper simultaneously studied the disappearance rate of four organophosphates on fruits and leaves.[36] They found that fruit and leaf surface residues were quite similar and suggested only one be analyzed for residues — preferably leaves. The disappearance rates of the four tested compounds are arranged in a decreasing order: malathion, oxydemeton methyl, dialifor, and dioxathion. The authors suggested that physical properties such as the diffusion coefficient and/or vapor pressure, which are temperature-dependent, may account for the observed differences in disappearance rates.

Nigg et al. studied the disappearance rates of acephate, methamidophos, and malathion for citrus foliage.[37] Malathion had a half-life of 5.2 d, acephate — 6.8 d (mist blowers) and 12.6 d (hand gun), and methamidophos — 9.5 d (mist blowers) and 13.6 d (hand gun). Contact with citrus foliage 7 or more days after application was not deemed dangerous.

It was stressed that only dislodgable residue data but not organic solvent extraction data should be used to estimate worker exposure. With special, simple equipment, small disks are taken from the foliage and analyzed.[21,35,37]

The relationship between leaf residues and dermal contamination was studied. Popendorf et al. calculated that 98 to 99% of the accepted dose in exposed persons is from dermal exposure, especially through the hands.[8] There was a correlation between calculated dose and the leaf residues of the pesticide.

Nigg et al. found a strong positive correlation between leaf pesticide residue and pesticide flux ($\mu g/cm^2/h$) onto various exposure pads, except shoulder, upper arm, and forearm pads.[24] The correlation coefficient for upper total exposure vs. leaf residue was 0.70; hand exposure vs. leaf residue was 0.97, and lower body exposure vs. leaf residue was 0.98. Only 1.6% of the estimated quantity of pesticide reaching the pad was excreted in urine.

Zweig et al. considered the relationship between the dermal exposure (dose) and the presence of pesticide on the leaves.[38] The dermal exposure was measured in mg/h. These authors found a positive correlation between these two parameters of exposure and pesticides of different groups: captan, vinclosolin, methiocarb, and carbaryl applied to different cultures. The relationship between dermal exposure and residual quantity on the leaf was expressed by the surface/time relationship. The ratio dermal exposure: dislodgable foliar residues was within the same order of magnitude for different pesticides and crops.

Skin exposure on the hands is estimated from cotton gloves, and on other parts of the body from the lint layer on the clothes. The pesticide content of the leaf is determined from 3 cm diameter disks cut out from the leaf.

Dermal exposure in mg/h is calculated by multiplying the amount of dislodgable foliar residues of leaf surface in mg/cm^2 by an empirically found ratio of approximately 5×10^3.

Calculating reentry intervals according to EPA guidelines goes through several steps. The noneffect level in $\mu g/kg/d$ is determined by calculating dermal dose — ChE response curve on rats.[34]

The allowable exposure level (AEL) is calculated from the NOELs for the pesticide using appropriate safety factor. It is assumed that 100% of the dose is absorbed during the 8-h working day of a 70-kg individual.

NOEL includes the results of dislodgable residues, the whole body dermal dose, and a dislodgable residues dissipation curve, obtained under field conditions.

Nigg and Stamper published an overview of field methods for assessing worker exposure.[23] Specific attention was given to dermal absorption pads and the statistical variability of their contamination, personal air samplers, and urine analysis. It was underlined that relevant human biochemistry and physiology data, as well as urinary excretion kinetics, could be a very worthwhile prior investment.

Mathematical equations for calculating reentry intervals are proposed in the U.S.S.R.[28]

III. STUDIES ON INDIVIDUAL PESTICIDES

Zweig et al. studied the dermal exposure of strawberry harvesters to benomyl and captan.[39] The ratio of the dermal amount of both pesticides was found to be similar to the ratio for the dislodgable foliar residues of the same two pesticides. The average dermal exposure was found to be 39 mg/h/person for captan and 5.39 mg/h/person for benomyl. Productivity as measured by the quantity of fruits picked, and the dermal exposure of the individuals to benomyl correlated positively; no correlation was found for captan. Similar results have been obtained for carbaryl.[39]

Zweig et al. studied the dermal exposure of strawberry harvesters to carbaryl.[40] They estimated the carbaryl exposure from cotton gauze, light cotton gloves, and plants. The foliar residues were 7.74 $\mu g/cm^2$ 1 d after application and 0.15 $\mu g/cm^2$ 17 d after application. The dermal hand exposure, at 3 mg/h on the first day, was the largest; on the third day it was 1.47 mg/h. With zolon in doses from 10 to 14 mg/h, no erythrocyte ChEA inhibition was provoked in pear harvesters.

Variations in the decay and oxidation of parathion orange foliar residues were related to ambient weather and foliar dust. Peak levels of parathion generally occur within 2 to 3 d.[41,42] The ozone concentration in the air also contributes to the conversion of parathion to paraoxon.[43] Compared to its production at low dust levels, more paraoxon was produced by a factor of 4 at high dust levels without ozone and by nearly a factor 30 at high dust levels in the presence of 300 ppb of ozone.

Reentry hazards and safe reentry periods in captan treated fields were also studied in the last few years.[38,43-45] Captan and its major metabolite, tetrahydrophthalimide (THPI), were determined as dislodgable residues on foliage and fruits, as well as in air samples and patches attached to clothing and gloves. As a biological index, the urine of workers was examined for the presence of tetrahydrophthalimide. THPI was found in the urine of field workers not protected by aspirators, but in spite of the high dermal exposure it was absent in the urine of those equipped with respirators. Respiratory exposure is believed to be the primary route. A study of workers in captan treated grape fields reached the same conclusion.[45]

A study by Batista et al. of a Florida citrus plantation found the half-life for carbaphenthion dissipation from leaves was about 4 d; the dissipation from soil was about 7 d.[46]

In the urine of 4 male workers harvesting onions in an ethyl- and methylparathion-treated field, traces or levels above the detection limit of diethyl phosphate and diethylthiophosphate were found 7 weeks after treatment.[3]

IV. REENTRY INTERVALS FOR SOME PESTICIDES

The state of California established occupational reentry safety standards for citrus fruit, peaches, grapes, and apple. The EPA has since developed standards, but these are less stringent than the California standards.[29]

The norms existing in U.S. are very different. For 12 pesticides the EPA has determined a maximum interval of 48 h or the drying up of the sprayed pesticide if corresponding reentry periods do not exist. California has established reentry intervals from 1 to 45 d for 20 pesticides. For example, the EPA standard for azinphosmethyl is 14 to 24 h and California standard is 14 to 30 d. Similar to our observations, they also give different terms depending on the concentration of the applied pesticide, e.g., parathion has from 30 to 60 d on citrus cultures.[29]

This fact as well as the possibility of some pesticides transforming into more toxic metabolites in the environment (e.g. parathion into paraoxon) requires investigation to determine reentry intervals.[30,47]

The largest reentry period proposed is 70 d for chlorthiophos in California citrus trees.[34,48]

Carbamates also represent reentry hazards. Studies on reentry intervals for carbamates have been published.[39,48,49]

Iwata et al. determined a safe dislodgable foliar level for carbosulfan and carbofuran — 0.3 μg/m^3 of leaf surfaces.[48] They conducted dose-ChE response studies on rats, and proposed a 7 d reentry interval for California citrus.

Using the same methodology, Nigg et al. proposed safe reentry intervals for Florida citrus groves treated with carbosulfan — 3 d for 4 lb of active ingredient/acre and 1 to 2 d for 1 lb of active ingredient/acre.[49]

For some pesticides, under specific conditions, no reentry intervals appeared necessary. Such is the case with difocol and deltamethrin applied to lemon trees and French beans, respectively.[47,50]

The reentry periods accepted in Bulgaria are shown in Table 1.

V. THE PROBLEM OF PRACTICAL USE OF REENTRY INTERVALS

U.S. reentry intervals are reglemented by local and federal agencies.[54-56]

Culver criticized the concept of worker reentry intervals as unrealistic in the long run and not

TABLE 1
Reentry Intervals in Bulgaria (in Days)[51-53]

Pesticide	Field	Glass houses
Dichlorphos	—	6
Fosdrin	—	3
Methylparathion	10	—
Tamaron	8	—
Vidate	—	4
Dimethoate	7[a]–5[b]	3
Intrathion	15[a]–13[b]	—
Sayfos	13[a]–11[b]	—
Actelik	—	1
Pirimor	5[a]	5
Unden	—	3

[a] Application by tractor.
[b] Aerial application.

a permanent solution.[57] As an alternative, he suggested preventive medical programs, which should include a variety of laboratory tests to ensure no negative health effect in the working environment. The supervising physician should remove workers from the field until the source of the problem is identified and corrected; they should remain off the field until their health has recovered. It is obvious that this methodology does not ensure real prevention in certain cases involving very toxic substances and early reentering. The authors do not discuss the alternative of avoiding highly toxic substances when there is a necessity to enter the field several days after treatment.

Reentry intervals are difficult to implement because of the inherent complexity of modern pest control technology, i.e., multiple chemicals, formulations, and dosage rates, as well as unpredictable climate conditions that affect the degradation of residues. However, extensive research, to identify safe reentry intervals for a large number of pesticides applied to a great variety of crops, is justified.

If this period can not be observed, persons working in sprayed areas must be warned of the risk and at least provided with protective gloves.

Regulations must require posted signs at reasonable intervals along the periphery of treated fields or orchards. The signs must warn against premature entry and state the date, prior to which, entry is illegal.

VI. CONCLUSIONS

Reentry intoxication may occur when workers enter previously treated fields for further cultivation and harvesting. The reentry interval is the time needed for pesticides on plants to degrade to a no effect level. The methodology of determining such periods includes different approaches, the principals being: the disappearance rate of fruit and leaf surface residues, dermal exposure, and dermal dose — the cholinesterase response curve on humans or experimental animals.

Reentry periods are difficult to establish and implement because of the complexity of pest control technology and changing climatic conditions. Nevertheless, research in this field is justified because of the large population entering treated fields.

REFERENCES

1. **Kaloyanova-Simeonova, F.,** Prevention of intoxication by pesticides in Bulgaria, *Whither Rural Medicine* Proc. 4th Int. Congr. Rural Med., Usuda (Tokoyo, 1970, 35–37.
2. **Griffith, J. and Duncan, R. C.,** Grower reported pesticide poisonings among Florida citrus fieldworkers, *J. Environ. Sci. Health, B.* 20, 1, 61, 1985.
3. **Munn, S., Keefe, T. J., and Savage, E. P.,** Comparative study of pesticide exposures in adult and youth migrant field workers, *Arch. Environ. Health,* 215, 40, 4, 1984.
4. **Kaloyanova, F.,** *Pesticides. Toxic Action and Prophylaxis,* Bulgarian Academy of Science, Sofia, 1977.
5. **Spear, R. C., Jenkins, D. L., and Milby, T. H.,** Pesticide residues and field workers, *Environ. Sci. Technol.,* 9, 4, 308, 1975.
6. **McClure, C. D.,** Public health concerns in the exposure of grape pickers to high pesticide residues in Madera County, Calif., Public Health Report, September, 1976.
7. **Peoples, S. A. and Maddy, K. T.,** Organophosphate pesticide poisoning, *West. J. Med.,* 129, 4, 273, 1978.
8. **Popendorf, W., Spear, R., Leifingwell, J., Yager, J., and Kahn, E.,** Harvester exposure to zolone (Phozalone) residues in peach orchards, *J. Occup. Med.,* 21, 3, 189, 1979.
9. **Kundiev, J. I., Nikitin, D. P., and Hoholkova, G. A.,** New hygienic problems related to large use of chemicals in agriculture, *Gig. Sanit.,* 10, 6, 1975 (in Russian).
10. **Klissenko, M. A., Viotenko, I. G., and Kisseleva, N.,** Toxicological and chemical investigations on volatile compounds forming in the soil containing polychlorpinen and mineral fertilizers, *Gig. Tr. Prof. Zabol.,* 10, 32, 1977 (in Russian).
11. **Gromov, V. L. and Gromova, V. S.,** Effect of pesticides on the working conditions in the weeding of sugar beet plantations, *Gig. Tr. Prof. Zabol.,* 11, 43, 1977, (in Russian).
12. **Van den Oever, R., Roosels, D., and Lahaye, D.,** Acute hazard of methyl bromide fumigation in soil desinfection, *Br. J. Ind. Med.,* 39, 140, 1980.
13. **Izmirova, N., Benchev, I., Kaloyanova, F., Kostova, S., and Angelova, R.,** Determination of minimum reentry intervals after tobacco spraying by sayfos, 5th Int. Congr. of Rural Med., Varna, (Abstracts), 1972, 122.
14. **Izmirova, N., Benchev, I., Velichkova, V., Kaloyanova, F., Lambreva, E., Russeva, P., Kostova, E., Angelova, R., Inkova, M., Bojinova, L., Maximova, F., and Tsvetkova, P.,** Determination of minimal reentry intervals after tobacco and cotton treatment by organophosphorous pesticides, *Gig. Primen. Toksikol. Pestits. Klin. Otravlenii,* 181, 1973, (in Russian).
15. **Izmirova, N., Kaloyanova, F., Kalinova, G., and Simeonov, P.,** New approach to hygienic standardization of pesticides, *Probl. Hig.,* 13, 98, 1988.
16. **Gutrie, F., Domanski, J., Main, A., Sanders, D., and Monroe, R.,** Use of mice for initial approximation of reentry intervals into pesticide treated fields, *Arch. Environ. Contam. Toxic.,* 2, 3, 233, 1974.
17. **Kraus, J. F., Mull, R., Kurtz, P., Winterlin, W., Franti, C. E., Kilgore, W., and Borhani, N. O.,** Monitoring of grape harvesters for evidence of cholinesterase inhibition, *J. Toxicol. Environ. Health,* 7, 1, 19, 1981.
18. **Spear, R., Popendorf, W., Leftingwell, J., Milby, T., Davies, J., and Spencer, W.,** Field workers response to weathered residues of parathion, *J. Occup. Med.,* 19, 6, 1977.
19. **Popendorf, W. J. and Leftingwell, J.,** Regulating OP pesticide residues for farmworker protection, *Residue Rev.,* 82, 125, 1982.
20. **Popendorf, W. J.,** Advances in the unified field model for reentry hazards, in Proc. Symp. on Risk Assessment of Agricultural Field Workers Due to Pesticide Dermal Exposure, American Chem. Society, 1985, 323.
21. **Nigg, H. N., Albrige, H. E., Nordby, H. E., and Stamper, J. H.,** A method for estimating leaf compartmentalization of pesticides in citrus, *J. Agric. Food Chem.,* 29, 750, 1981.
22. **Nigg, H. N. and Stamper, J. H.,** Considerations in worker reentry, AGS: Symposium Series No. 182, Pesticide residues and exposure, Plimmer, J. K., Ed., 1982, 59.
23. **Nigg, H. N. and Stamper, J. H.,** Field studies: Methods overview, Proc. Symp. on Risk Assessment of Agricultural Field Workers Due to Pesticide Dermal Exposure, American Chem. Society, 1985, 95.
24. **Nigg, H. N., Stamper, J. H., and Queen, R. M.,** The development and use of a universal model to predict three crop harvester pesticide exposure, *Amer. Ind. Hyg. Assoc. J.,* 45, 3, 182, 1984.
25. **Kahn, E.,** Outlines of guide for performance of field studies to establish reentry intervals for organophosphate pesticides, *Residues Rev.,* 70, 27, 1979.
26. **Skinner, C. S. and Kligore, W. W.,** Application of a dermal self-exposure model to worker reentry, *J. Toxicol. Environ. Health,* 9, 3, 461, 1982.
27. **Hugg, H.,** Prediction of agricultural worker safety reentry times for organophosphate insecticides, *Am. Ind. Hyg. Assoc. J.,* 41, 340, 1980.
28. **Spinu, E. I., Bolotnyj, A. V., Zoreva, T. D., and Ivanova, L. N.,** Determining safe waiting times prior to entry into zones treated with pesticides, *Gig. Sanit.,* 11, 61, 1980.
29. **Milby, T. H.,** Ed., Occupational exposure to pesticides, January 1974, Federal Working Group on Pest Management, Washington, D.C., 1974.

30. **Nigg, H. N.,** Prediction of agricultural worker safety reentry times for organophosphate insecticides, *Am. Ind. Hyg. Assoc. J.,* 41, 5, 340, 1980.

31. **Wicker, G. W. and Guthrie, F. E.,** Worker crop contact analysis as a means of evaluating reentry hazards, *Bull. Environ. Contam. Toxicol.,* 24, 1, 161, 1980.

32. **Stamper, S. H., Nigg, H. N., and Winterir, W.,** Growth and dissipation of pesticide oxon, *Bull. Environ. Contam. Toxicol.,* 27, 512, 1981.

33. **Spear, R. W., Popendorf, W., Spencer, W., and Malby, T.,** Worker poisoning due to paraoxon residues, *J. Occup. Med.,* 19, 6, 411, 1977.

34. **Iwata, Y., Knaak, J. B., Carman, G. E., Düsch, M. E., and Gunther, F. A.,** Fruit residue data and worker reentry research for chlorthiophos, applied to California citrus, trees, *J. Agric. Food Chem.,* 30, 215, 1982.

35. **Nigg, H. N. and Stamper, J. H.,** Dislodgeable residues of chlorbenzilate in Florida citrus: worker reentry implications, *Chemosphera,* 13, 10, 1143, 1984.

36. **Nigg, H. N. and Stamper, J. H.,** Comparative disappearance of dioxathion, malathion, oxydemeton methyl and dialifor from Florida Citrus Leaf and Fruit Services, *Arch. Environ. Contam. Toxicol.,* 10, 497, 1981.

37. **Nigg, H. N., Reiner, J. A., Stamper, J. H., and Fitzpatrick, G. E.,** Disappearance of acephate methamidophos and malathion from citrus foliage, *Bull. Environ. Contam. Toxicol.,* 26, 267, 1981.

38. **Zweig, G., Leffingwell, J. T., and Popendorf, W.,** The relationship between dermal pesticide exposure by fruit harvesters and dislodgeable foliar residues, *J. Environ. Sci. Health, B.,* 20, 1, 27, 1985.

39. **Zweig, G., Gao, R., and Popendorf, W.,** Simultaneous dermal exposure to captan and benomyl by strawberry harvesters, *J. Agric. Food Chem.,* 31, 1109, 1983.

40. **Zweig, G., Gao, R., Witt, J. M., Popendorf, W., and Bogen, K.,** Dermal exposure to carbaryl by strawberry harvesters, *J. Agric. Food Chem.,* 32, 1232, 1984.

41. **Popendorf, W. J. and Leffingwell, J. T.,** Natural variation in the decay and oxidation of parathion foliar residues, *J. Agric. Food Chem.,* 26, 2, 437, 1978.

42. **Popendorf, W.,** Exploring citrus harvesters exposure to pesticide contaminated foliar dust, *Am. Ind. Hyg. Assoc. J.,* 41, 652, 1980.

43. **Spear, R. C., Lee, Y. S., Leffingwell, J. T., and Jenkins, D.,** Conversion of parathion to paraoxon in foliar residues: effect of dust level and ozone concentrations, *J. Agric. Food Chem.,* 26, 2, 434, 1978.

44. **Winterlin, W. L., Kilgore, W. W., Mourer, C. R., and Sarah, R. S.,** Worker reentry study for captan applied to strawberries in California, *J. Agric. Food Chem.,* 32, 664, 1984.

45. **Winterlin, W. L., Kilgore, W. W., Mourer, R., Hall, J., and Hadapp, D.,** Worker reentry into captan-treated grape fields in California, *Arch. Environ. Contam. Toxicol.,* 15, 3, 301, 1986.

46. **Batista, G. C., Stamper, J. H., Nigg, H. N., and Knapp, J. I.,** Dislodgeable residues of carbophenothion in Florida citrus: implications for safe worker reentry, *Bull Environ. Contam. Toxicol.,* 35, 213, 1983.

47. **Staiff, D. C., Davies, J. E., and Robbins, I. L.,** Parathion residues on apple and peach foliage as affected by the presence of the fungicides maneb and zineb, *Bull. Environ. Contam. Toxicol.,* 17, 3, 253, 1977.

48. **Iwata, Y., Knaak, J. B., Düsch, M. E., O'Neal, J. R., and Pappae, J. L.,** Worker reentry research for carbosulfan, applied to California citrus trees, *J. Agric. Food Chem.,* 31, 1131, 1982.

49. **Nigg, H. N., Stamper, J. H., Knaak, J. B.,** Leaf, fruit and soil surface residues of carbosulfan and its metabolites in Florida citrus groves, *J. Agric. Food Chem.,* 32, 80, 1984.

50. **Mestres, R., Francois, Causse, C., Vian, L., and Winnett, G.,** Survey of exposure to pesticides in greenhouses, *Bull. Environ. Contam. Toxicol.,* 35, 6, 750, 1985.

51. **Kaloyanova, F. and Izmirova, N.,** Prevention in pesticide application in *Manual of Safe Use of Pesticides,* Kaloyanova, F., Ed., Medicina i Fizkultura, Sofia, 1985, 7 (in Bulgarian).

52. **Kaloyanova-Simeonova, F., Izmirova-Mosheva, N.,** Determination of minimum periods for safe work following spraying with organophosphate pesticides, in *IUPAC Pesticide Chemistry. Human Welfare and the Environment,* Pergamon Press, Oxford, 1983, 237.

53. **Kalinova, T. N.,** Determination of Minimal Reentry Intervals for Safe Work with Some Organophosphate and Carbamate Pesticides, Ph.D. thesis, Sofia, 1987.

54. Federal Register, 1975, 40, 26900-26901.

55. U.S. Environmental Protection Agency, Unpublished proposed regulations, 1981.

56. **Knaak, J. B.,** Minimizing occupational exposure to pesticides: techniques for establishing safe levels of foliar residues, *Residue Rev.,* 75, 81, 1980.

57. **Culver, B. D.,** Worker reentry safety. VI. *Occupational Health Aspects of Exposure to Pesticide Residues,* Springer-Verlag, New York, 1976, 41.

INDEX

A

Abate
  dose-effect relationship, 8
  permissible levels, 32
Acephate, 182
Acephatex, 182
Acetophos methyl, 32
Acrex, 132
Actelik, 185
Acute intoxication, epidemiology of, 173–179
Acylate, 54
Afugan, 32
Agent Orange, 142
Aldicarb, 44
  acute intoxication, 47
  epidemiology of acute intoxication, 176
  mechanisms of action, 46
  metabolism, 43, 45
  permissible levels, 53
  volunteer studies, 48
Aldoxycarb, 43
Aldrin, 59
  acute intoxication, 77
  carcinogenicity, 65
  chronic effects, 86–89
  LD_{50}, 63
  levels in workers, 76, 77
  percutaneous penetration, 60
  permissible levels, 90, 91
  physical properties, 62
  protein binding, 65
  structure, 61
Alkylcarbamates, occupational exposure, 50
Alkylcarbamyls, ChE reactivation time, 45
Alkylmercury compounds, 117–122
Alkylparathion, reentry intoxications, 184
Alkylphenylcarbamates, 54
Alkyl-thiophanate, 53
Alkyltin compounds, see Organotin compounds
Allethrin, 102, 103, 105
Allethroline, 101
Allycinerin, 102
Allylxycarb, 43
Alphamethrin, 101, 103
Ambush, 102
Ambushfog, 102
Amidithion, 32
Amidophos, 32
Aminocarb, 43
  mechanisms of action, 46
  metabolism, 43
  permissible levels, 53
Amiton, 4
Amitrol, 163
Antagonisms, combined pesticide exposure, 161
Anthio, 21, 32
Antipyrin, 81
Aquacide (diquat), see Dipyridiliums
Aquatin (triphenyltin chloride), 112
Aremo, 103
Arseniacals, 174

Asulam, 44
Axothoate, 4
Azinphos ethyl, 4
Azinphos methyl (gluthion), 4, 181
  combined pesticide exposure, 163
  dose-effect relationship, 8
  permissible levels, 32
  monitoring exposure, 23, 24, 25
  reentry intoxications, 184
Azocyclotin, 112
Azodrin, 175

B

Baricard, 103
Bastox, 103
Basudin, 32
Bavistin (carbendazim), 131, 132
Baygon, see Propoxur
Baytex, 32
2,4-B [(2,4-dichlorophenoxy)butyric acid], 136
Belmark, 103
Bendiocarb, 43
  metabolites, 45
  permissible levels, 53
Benlate, see Benomyl
Benomyl, 131, 132
  combined pesticide exposure, 163
  exposure, 46
  permissible levels, 53, 54
  reentry intoxications, 183
  structure, 44
Benzene hexachloride (BHC), 59, 161
  chronic effects, 79
  epidemiology of acute intoxication, 177
  tissue concentrations, 71
    blood and urine, 72
    fat tissue, 67–69
    fetal tissues and newborns, 73
    pregnant women, 70, 74
Benzimidazole compounds, 44, 54, 131–133
Benzofuroline, 102
Betamal, 54
Betonal (fendifam), 54
Bidrin, 22
Bioallethrin, 101
Biobenzylfuroline, 102
p-Biomophenoxy triethyl tin, 113
Biopermethrin, 102
Bioresmethrin, 102, 103
Bipyridiliums, 177
BPMC, 43
Bretdan (fentin acetate), 112
Bromophos, 4
  dose-effect relationship, 8, 9
  permissible levels, 32, 33
Bromophos ethyl, 4
Bromoxyl, 161
Bromoxynil, 160
Bufencarb, 43
Butacarb, 43
Butontate, 5

K

Kafil, 102
Karate, 103
Katalon (paraquat), see Dipyridiliums
Kelthane, 59, 63
Kilval, 11, 32
Kothrin, 103

L

Lanate, 174
Landrin, 43
Lannate, 174, 175
Leptophos, 5, 11
   epidemiology of acute intoxication, 174
   neuropathy, delayed, 15
Lindane, 82
   acute intoxication, 78
   chronic effects, 85, 86
   exposure, 65
   in fat tissue, 67–69
   LD_{50}, 63
   percutaneous penetration, 60
   permissible levels, 90, 91
   physical properties, 62
   structure, 61

M

M-81 (*o,o*-dimethyl-*S*-ethylmercatproethyl-dithio-
    phosphate), 82
Malathion, 4
   acute nitoxication, 12
   cholinesterase inhibition, 27
   combined pesticide exposure, 162
   contact dermatitis, 11
   dose-effect relationship, 8, 9
   effects, 21
   entry routes, 9
   epidemiology of acute intoxication, 176
   metabolism, 3–7
   neuropathy, delayed, 16
   neurotoxicity, 14
   permissible levels, 32, 33
   reentry intoxications, 182
Maneb, 125–128
MCBA (4-Chloro-2-methyl phenoxybutyric acid), 136
MCPA (4-chloro-2-methyl phenoxyacetic acid), 136
   effects, 139, 140
   exposure, 138, 139
MCPA (4-chloro-2-methyl phenoxyacetic acid), 136
MCPP [α-4-Chloro-2-methyl phenoxy)
    propionic acid], 136
Mecarbam, 4
Mecoprop [α-4-Chloro-2-methyl phenoxy)
    propionic acid], 136
Menazone, 4
Meothrin, 103
Merbate, 48
Mercaptophos, 13
Mercury compounds, 117–122
Merphos, 14–16
Metacrifos, 33
Metam dosium (vapam), 125, 126
Metamidophos, see Tamaron
Metasistox, 20, 21

Methafos, 32
Methamidofos, 4
   neuropathy, delayed, 15
   permissible levels, 33
   reentry intoxications, 182
Methamyl, 46
Methidothion, 4
   dose-effect relationship, 8
   permissible levels, 33
Methiocarb, 44
   mechanisms of action, 46
   permissible levels, 53
   reentry intoxications, 183
Methomyl, 44
Methoxychlor, 59
   LD_{50}, 63
   permissible levels, 91
   physical properties, 62
   structure, 61
Methyl-2-benzimidazole carbamate, 131
Methylbromide, 174
Methylcarbamyl, 45
Methyldemeton, 23
Methylethylphos, 13
Methylmercaptophos (sistox), 20, 21, 32
Methyltin, 111
Methyl topsin (thiophanate methyl), 131, 132
Metrifonate, 24
Mevinphos, 4
   cholinesterase inhibition, 27
   combined effects, 162
   dose-effect relationship, 8
   effects, 22
   EMG findings, 16
   monitoring exposure, 23
   permissible levels, 32, 33
Mexacarbate, 44, 47
Mipaphox, 15, 27
Mirex, 68
Monocrotophos, 4, 32, 33
Morphothion, 4

N

Nabam, 125, 126
Naled, 4, 32
Necarboxyl acid, 102
Neopynamin, 102
Nicotine, 174, 175
Nitrophenols
   combined pesticide exposure, 166
   epidemiology of acute intoxication, 174
Nitrosocarbaryl, 54
Nizole, 160

O

Obidoxime, 31
Octachlor, 61
Octamethyl, 32
Omethoate, 4
   neurotoxicity, 14, 16
   permissible levels, 33
op′DDD, 64
Organochlorine compounds
   acute effects, 77, 78

DUE DATE

APR. 0 4 1993			
MAY 1 0 1993			
OCT 0 4 '98			
OCT 0 '98			
OCT 0 4 '98 ILL: 4954456 AD#			
MAY 2 0 2009			
			Printed in USA